Interstate Water Resource Risk Management *Towards a Sustainable Future for the Aral Basin* (JAYHUN)

Interstate Water Resource Risk Management *Towards a Sustainable Future for the Aral Basin* (JAYHUN)

Edited by

Oliver Olsson and Melanie Bauer

IWA Publishing
London • New York

Published by **IWA Publishing**
Alliance House
12 Caxton Street
London SW1H 0QS, UK
Telephone: +44 (0)20 7654 5500
Fax: +44 (0)20 7654 5555
Email: publications@iwap.co.uk
Web: www.iwapublishing.com

First published 2010

Typeset in India by OKS Prepress Services.

British Library Cataloguing in Publication Data
A CIP catalogue record for this book is available from the British Library

Library of Congress Cataloging-in-Publication Data
A catalog record for this book is available from the Library of Congress

ISBN 10: 1843393085
ISBN 13: 9781843393085

Contents

 Interstate Water Resource Risk Management *Towards a Sustainable Future for the Aral Basin* (JAYHUN). Edited by Oliver Olsson and Melanie Bauer. ISBN: 9781843393085. Published by IWA Publishing, London, UK.

Preface

Water – the essence of life and possibly the most important good in the regions of Central Asia – is becoming increasingly limited. Available water resources are decreasing further and further due to the existing impact of climate change and due to problems arising from mismanagement and pollution.

Glacial and snow melt is essential for the well being of all central Asian states, as this provides over 90% of the water demand. Climate change is causing the rapid recession of glaciers, which in the short term helps these states to meet their ambitious water requirements, but in the long term will result in a decreased runoff and increased evapotranspiration from higher temperatures. Moreover, effects such as high levels of erosion from the young and unstable mountains generate a rapid siltation of these important water supply reservoirs, thus limiting their life expectancy.

This book is developed by the Jayhun Consortium under the EU-INCO FP6 research project Jayhun: Interstate Water Resource Risk Management: Towards a Sustainable Future for the Aral Basin. It focuses on and discusses the results of research done on the risk assessment levels associated with regional relevant variables and to develop a strategy for an improved management of the region's water resources.

The book provides recent and reliable information on the rate of glacial retreat and the implications for future water resources, it was possible to assess the impact of global changes on the annual precipitation in the region. The rates of reservoir siltation were also established by hydrographic surveys, and analyses of wastage and effectiveness of the water usage in the large irrigation massifs were conducted based on historic data and methods involving satellite data. Existing and newly generated data were analysed using water resources models. These were calibrated with existing data and subsequently used for simulating different scenarios up to 80 years into the future.

The central and most important result shown in this book is a risk analysis on future risk and the identification of sustainable management options for the water resources of the Aral Basin.

List of Contributors

Agaltseva, Natalya
Hydrometeorological Research Institute,
Uzhydromet, 72, K.Makhsumov,
700052 Tashkent, Uzbekistan
e-mail: natagalt@mail.ru

Bauer, Melanie
Water Resources Management Division,
Institute for Water Quality and Waste Management,
Leibniz Universität Hannover, Callinstr. 14,
30167 Hannover, Germany
e-mail: mb@warb.uni-hannover.de

Bolgov, Mikhaïl
Water Problem Institute,
Russian Academy of Sciences,
GSP-1, Gubkin St., 3, IWP RAN,
119997 Moscow, Russian Federation
e-mail: bolgov@aqua.laser.ru

Chevallier, Pierre
Laboratoire Hydrosciences
(CNRS, IRD, Montpellier 1 et 2 Universities),
Maison des Sciences de l'Eau, CC 57,
34095 Montpellier Cedex 5, France
e-mail: pierre.chevallier@ird.fr

Froebrich, Jochen
Centre for Water and Climate, Alterra,
P.O. Box 47, 6700 AA Wageningen,
The Netherlands
e-mail: jochen.froebrich@wur.nl

Ikramova, Malika
Central Asian Scientific Research Institute
of Irrigation (SANIIRI), Karasu – 4, 11/41,
700187 Tashkent, Uzbekistan
e-mail: dilmalik@tps.uz

Khodjiev, Alisher
Central Asian Scientific Research Institute
of Irrigation (SANIIRI), Karasu – 4, 11/41,
700187 Tashkent, Uzbekistan
e-mail: dilmalik@tps.uz

Lisenko, Oleg
Amudarya Basin Water Resources
Management Organisation,
63, Zamahshari, Urgench,
Uzbekistan 74000
e-mail: olelysenko@yandex.ru

Mojaïsky, Marie
Association VERSeau Développement,
859, rue Jean-François Breton,
34093 Montpellier Cedex 5, France
e-mail: mojaisky@verseaudeveloppement.com

Normatov, Inom
Institute of Water Problems,
Hydropower and Ecology,
Academy of Sciences, Parvin Str.12,
734002 Dushanbe, Tajikistan
e-mail: inomnor@mail.ru

Olsson, Oliver
Water Resources Management Division,
Institute for Water Quality and Waste Management,
Leibniz Universität Hannover, Callinstr. 14,
30167 Hannover, Germany
e-mail: oo@warb.uni-hannover.de

Pouyaud, Bernard
Association VERSeau Développement,
859, rue Jean-François Breton,
34093 Montpellier Cedex 5, France
e-mail: bernard.pouyaud@free.fr

Rycroft, David W.
School of Civil Engineering & the Environment,
University of Southampton, Highfield,
Southampton SO17 1BJ, UK
e-mail: dwr@soton.ac.uk

Sharipov, Ilkhom
Bathymetric Centre, 1. Usman Usupova,
Tashkent, 7000125, Uzbekistan
e-mail: baxtiyorxoja@yahoo.com

Sorokin, Anatoly
Scientific Technical Centre Clean River,
"Toza Darya", Karasu – 4, 11/41, 700187
Tashkent, Uzbekistan
e-mail: sorant@mail.ru

Spectorman, Tatjana
Hydrometeorological Research Institute,
Uzhydromet, 72, K. Makhsumov,
700052 Tashkent, Uzbekistan
e-mail: mdspectr@glb.net

Tanton, Trevor W.
School of Civil Engineering & the Environment,
University of Southampton, Highfield,
Southampton SO17 1BJ, UK
e-mail: t.w.tanton@soton.ac.uk

Trubetskova, Marina
Water Problem Institute,
Russian Academy of Sciences,
GSP-1, Gubkin St., 3, IWP RAN,
119997 Moscow, Russian Federation
e-mail: trubets@mail.ru

Vauchel, Philippe
Laboratoire Hydrosciences (CNRS, IRD,
Montpellier 1 et 2 Universities),
Maison des Sciences de l'Eau, CC 57,
34095 Montpellier Cedex 5, France

White, Christopher
School of Civil Engineering & the Environment,
University of Southampton, Highfield,
Southampton SO17 1BJ, UK

Chapter 1

A look forward to the management on the water resources of Central Asia

Oliver Olsson, Melanie Bauer and Jochen Froebrich

1.1 WATER RESOURCES MANAGEMENT IN THE AMU DARYA BASIN – NEED FOR A LOOK FORWARD

Reliable and safe supply of fresh water resources, together with a sound agricultural development and ecosystem preservation remains as one of the most important global environmental challenges. Increasing demand for water, as a result of a growing global population and their extensive use of water for irrigation and industry, has raised the awareness of our vulnerability to drought. Severe consequences of drought can be found in arid regions where the availability of water is already limited, a situation that is further aggravated when climate change might lead to an increased frequency and severity of droughts.

Besides many effects, climate change may strongly affect the coverage of snow and glaciers and moreover it impacts on the entire water cycle by changing precipitation pattern and water flow regimes. In which extent those influences and impacts are and will happen cannot be quantified precisely but already identified changes evident their existence.

Water resource management must become economically more efficient, ecologically sustainable, and also socially justifiable. It is especially so in the regions of the world suffering from water crisis (Martius *et al.*, 2007). The request for safe water supply in terms of water quality and quantity is also a core element of the Millennium Development Goals (MDGs), agreed by the world leaders at the United Nations Millennium Summit in 2002.

The Amu Darya River Basin in Central Asia is one of the crisis regions worldwide where the management of water resources is unsustainable, uneconomic and hampered by its cross-border extension. By the latter there is also a great potential for social conflicts. Central Asia is still an area of economic and political turmoil and change. Smith (1995), looking at water sharing arrangements, argues "nowhere in the world is the potential for conflict over the resources as strong as in Central Asia". Although this fact importantly affects the water resources management of the single independent states, focussing on those conflicts was beyond the scope of the project and will not be rated further in detail.

In this context a forward look for water resources management in the Amu Darya Basin is needed more than ever. For example, too long the debate had focused on the disappearance of the Aral Sea, discussing causes and impacts, while the extensive abstraction of water from Amu Darya and also Syr Darya is more than evident the origin of the Aral Sea problems. Furthermore, the discussion on the future use of land and water resources in Central Asia is certainly not new and have been addressed also in Tanton and Heaven (1999) or Martius *et al.* (2007). Thereby, less attention have been paid to the specific adaptation of management schemes to droughts and reduced river water availability during dry years. Work carried out within the IWMT project (INTAS-00-1043/DFG FR1607/1-1), Olsson *et al.* (2009) and

Wegerich (2009) indicated that irrigated agriculture and livelihood is hit already hardest during these dry years. With that and new insights on future climate and global changes require a contemporaneous consideration of current and future water shortages in the development of adaptation strategies.

1.2 KEY POINTS OF A LOOK FORWARD

A forward look on managing water resources in the Amu Darya basin require both the consideration of current water resources as well as the existence of an enormous water infrastructure.

Two major rivers flow into and feed the Aral Sea, the Amu Darya in the South, and the Syr Darya in the North (Table 1.1). The Amu Darya is the largest tributary in terms of run-off to the Aral Sea with a total mean discharge of 79.3 km^3 per year, with a high of 108 km^3 and a low of 47 km^3.

Table 1.1. Territorial annual renewable surface water resources in the Aral Sea basin.

Country or Zone	Renewable Surface Water Resources					
	Amu Darya basin		Syr Darya basin		Aral Sea basin	
	km^3 per year	% of basin	km^3 per year	% of basin	km^3 per year	% of basin
South-Kazakhstan	–	0.0	2.426	6.5	2.426	2.1
Turkmenistan	1.549	1.9	–	0.0	1.549	1.2
Uzbekistan	5.056	6.0	6.167	16.6	11.223	9.6
North-Afghanistan	21.593	27.2	–	0.0	21.593	18.6
Tajikistan	49.578	62.9	1.005	2.7	50.583	43.4
Kyrgyz Republic	1.604	2.0	27.605	74.2	29.209	25.1
Basin	79.280	100.0	37.203	100.0	116.483	100.0

The basin is divided into high mountain areas of the Pamir-Alai-System and desert areas of the Turan Plain that consists of the Kzylkum desert in the East and the Karakum desert in the West.

It is formed by the confluence of its main headwater tributaries, the Vakhsh and Pyanj Rivers. The total length of the Amu Darya from the head of the Pyanj River to the Aral Sea is about 2540 km, whereas the length from the river confluence accounts to 1415 km (Froebrich and Kayumov, 2004). The catchment area of the Amu Darya basin comprises 309,000 km^2 and is shared by Afghanistan and four Central Asia Republics: Kyrgyzstan, Tajikistan, Turkmenistan and Uzbekistan.

The Amu Darya tributary the Vakhsh River originates in the mountainous regions of the Pamir Alai in the southeast territory of Kyrgyzstan, where parts of the Abramov glacier and the Fedchenko glacier contribute to run-off generation. The Pyanj originates at the glacier in the Vakjdjir Pass, and forms the border between Afghanistan and Tajikistan. The tributaries of the Pyanj are located in Afghanistan and Tajikistan. The major tributaries from Afghanistan are the Wakhan, Pamir, Badkhshan and Kokcha. The major tributaries from Tajikistan on the Pyanj are Gunt, Bartang, Vanch and Kyzylsu. After the confluence of the Vakhsh and Pyanj, the Amu Darya receives water from the Kunduz (from Afghanistan), the Kafirnigan (from Tajikistan), the Sherabad and Surkhandarya (from Uzbekistan) rivers. The Afghan rivers Khulm, Balkh, Sar-e-Pul and Sherintang are mostly consumed locally and reach the Amu Darya only rarely (Ahmad and Wasiq, 2004). At Termez, the border city between Afghanistan and Uzbekistan, the river leaves the high mountains of Pamir, enters the desert plain of Karakum and then flows down to Turkmenistan. Upon returning to Uzbekistan the Amu Darya finally ends in the Delta downstream of Nukus in the Autonomous Republic of Karakalpakistan.

The hydrologic regime of the Amu Darya River can be described by an uneven temporal distribution of the annual runoff, with about 80% within the period April to September. High discharges occur twice a year, first for a short period in April/May after the snowmelt and second in June/July during the glacial melt in the Pamir-Alai Mountains. Hence, greatest amounts arise in mid summer. During the period from 1957 to 1980 the glaciers of the region have lost 126 km^3 of ice (around 113 km of water) caused by climate change, equally accounting for 19% of the total ice reserves in 1957 (National Commission of the Republic of Uzbekistan on Climate Change, 1999).

The Amu Darya basin is a typical endorheic basin under arid conditions. The climate of Uzbekistan is characterised by dry and hot summers due to the influence of local not continental air mass, and on the other hand by cold and unstable winter, formed by Arctic and Siberian air masses from the North. High temperatures have maximum values up to +50°C and minimum values down to −30°C. The annual precipitation in the lower Amu Darya region is only about 80 to 200 mm, concentrated in the winter and spring, while on the southern and southwestern sides of the Pamir-Alai Mountains it ranges between 600 and 800 mm and up to 1400 mm at high altitudes.

Most of the water of the Amu Darya derives from the high mountain glaciers of the Pamir-Alai-System, while the desert plains that cover about two thirds of the basin do not contribute significant amounts of runoff. In addition to no existing tributaries in the downstream part, high evaporation (Urgench, average: 1,364 mm/y), low precipitation (Urgench, average: 110 mm/y), and infiltration into groundwater cause a natural decrease of river discharge. High abstractions of water for irrigation, industrial and municipal supply, intensify the decline. Natural water shortage occurs in March, but due to the overexploitation of water resources in the river basin water shortages may occur also in spring and summer periods.

Agriculture without irrigation is not possible in Central Asia. For example, approximately 69% of the irrigated lands in Uzbekistan are located in the desert and semi desert zone, with annual rainfall varying from less than 97 mm in the northwest to 425 mm in the mountainous zone in the middle and southern parts of the country. The irrigated land has been steadily expanded from 4.5 million ha in 1960 to 8.2 million ha today. The annual agricultural water demand in the Aral Sea basin varied between 4,500 m^3/ha to 20,600 m^3/ha (1990 to 1998) but officially the water consumption is limited by interstate organizations up to 12,400 m^3/ha (ICWC, 2008).

Water allocation to all supplied sectors have been always an issue since the end of the Soviet Union. Moreover, there is an ongoing debate on the impact of large dams on the water availability and allocation within the Amu Darya basin. Unfortunately, until today both the regional and also the international discussion are constrained by speculations, assumptions and only sparsely available and poorly reliable data. After the break-up of the Soviet Union attention was first focused on the water sharing agreements and only later the attention shift to large dams as water control infrastructure in CA (Wegerich *et al.*, 2007). In addition to the issue of water resource allocation, the matter of water control infrastructure has become a point of contention. O'Hara (2000) interprets the building of dams and reservoirs upstream in CA as a political strategy. She reasons that under the Soviet Union the location had two advantages for Moscow and furthered its strategy of "divide and rule". Lange (2001), however, explains the dams in the Syr Darya basin in a non-political manner, stating, "the water management infrastructure was designed for a unified purpose and placed where it made sense". Hence, Lange sees the construction of dams upstream as a way to have full water control for downstream irrigated agriculture. According to this view, only downstream countries benefit. Similarly, Wegerich (2004) shows for the Syr Darya basin that the old and integrated management system with its issue linkage approach between water, energy and food benefited all the riparian states and offered a win-win solution. It is questionable whether any of these general arguments made for upstream water control infrastructure in Central Asia, or for the Syr Darya basin in particular, can be applied to the Amu Darya and the Nurek dam in upstream Tajikistan.

In terms of water allocation an additional major interstate conflict spot can be identified between Turkmenistan and Uzbekistan at the lower Amu Darya because of the Tuyamuyun Hydro-Complex. In comparison to the past Soviet system, especially the independency of Uzbekistan and Turkmenistan results in a much more complicated situation. Both countries have large irrigated areas and the question of regulating the multi-reservoir system is rather critical, and of strategic and political importance. The main part of the reservoir system is located on the territory of Turkmenistan, except the main dam with its operation unit and the main reservoir, which forms the border between Uzbekistan and Turkmenistan.

In terms of quantity, water should not be a scarce resource in the Aral Sea basin, not even in dry years. However, Central Asian countries inherited after their independence in 1991 a situation characterized by the heavily salinisised, polluted and impoverished soils, huge and partly dysfunctional irrigation systems, and a dependency on cotton as a cash crop. So main drivers limiting water resources are quality-related. Over the years for example land degradation caused among others by antiquated, not adapted irrigation techniques soil problems and adds further to the degradation of surface and groundwater resources. Hence future water stress and the impact of droughts is not affected by reduced water availability alone.

As part of the environment, water resources and their conditions are an important factor within the economy of a country. But furthermore four factors will drive mainly the intertwined environmental and economic development in the irrigated lands of Central Asia (Martius *et al.*, 2007): Demography, Climate change, Increased water demand and Land degradation.

Demography: Generally, there is a high population growth in Central Asia, particularly in Uzbekistan with a two per cent population growth rate (UNFCCC, 2001). About two thirds of the 44 million inhabitants of the Aral Sea basin depend on irrigated agriculture. Increased food demand due to high population growth rates will prompt further expansion of irrigated lands, whilst increasing the risk of water unavailability (MEA, 2005).

Climate change: Central Asia is experiencing a global warming above average, as exemplified by an average temperature gain of 1.2–2.1°C since the 1950s (far in excess of the 0.5°C increase in global temperature; Giese and Mossig, 2004), and a 1–2°C increase was observed during the last century in Central Asia, much higher than the regional average of 0.7°C (IPCC, 2001). The IPCC (2001) also projects a temperature increase of 1–2°C until 2030–2050 in the region.

In the context of climate change the impact of glacier melt on the run-off distribution, land – atmosphere interrelations and feedback or further mechanisms, like the increase of evapotranspiration, play an important role. As those needs are well established and elaborated by simulation models there is much to do for providing a sound basis for estimating climate change impacts on the states of Central Asia. Especially, the glacial and snowmelt is an essential source for the well being of all regions in Central Asia and provide over 90% of their water requirements.

The above-average climatic change predicted for Central Asia will bring short-term gains in water availability downstream, but will on a long-term scale translate into lower water inflow from the mountains, and higher water losses to the atmosphere. Additionally, climate change has an effect on the frequency and intensity of extreme droughts, with the consequence of increased exceptional water deficits as occurred for the lower Amu Darya River during 2000–2002 and 2008.

Increased water demand: The lowlands in the watershed are particularly vulnerable to increased water diversions upstream. The upstream countries (Tajikistan, Kyrgyzstan) will be gradually claiming more 'their' water resources for themselves. The lowland countries Uzbekistan and Turkmenistan consume 83% of all water of the basin. Turkmenistan is insisting on its share of the water resources, used in ambitious 'development' projects. Further demand will be created if Afghanistan proceeds with developing its irrigation systems at the Amu Darya, which may divert up to 10% of the total water budget of the region.

Land degradation: Land degradation is an economic, social, and environmental problem in the transition economies of all five CA countries. This effect seriously affects the livelihood of the rural population by reducing the productivity of the land and leading to agricultural losses as well as to increased costs for the agricultural production. The annual costs of land degradation have been estimated by the ADB (2004) at about 7 billion US Dollars for all five Central Asian countries.

1.2.1 Sustainable management of water resources

Everywhere in the world and especially in water scarce areas sustainability of water resources management is an issue par excellence. But at this, it is even more important to use already available opportunities and resources, infrastructures rather than building new constructions' or introducing new techniques. Coming back to the roots of water management schemes should be one of the aims in order to achieve the important sustainability.

The concept of a sustainable water resources management was first mentioned in Stockholm in 1972, during the United Nations World Conference, and then at the Rio summit in 1992 with Agenda 21 (Ganoulis, 2006). Following, current interest in sustainable water resources management has come from realising that some of the past or current activities could have caused irreversible damage to the ecosystem. This damage may adversely affect not only our own lives but also the lives of our successors. To address these questions, it is helpful to differentiate growth from development. According to Loucks *et al.* (2000), growth involves "making the pie bigger", building new capacity in new places, improving the standard of living, changing land use, etc. Development involves capacity expansion *in situ*, redistribution of existing resources, a more efficient use of scarce resources, water quality management, and the like.

Hence, understanding of sustainable development of water resources includes determination of existing and future demand of water resources, rational water use, comprehensive observation and assessment, effective supervision and provision of the necessary conditions for protection of related structure. Taking these into account, emphasis is given to supply water in sufficient quantity and quality at the right time and place, reallocating water according to certain priorities among sectors, conjunctive use of surface and groundwater resources and environmental factors.

The impact of drought must be seen as a dynamic process, basically as the result of interactions between water supply and demand. The relationship between those two aspects, the supply of water, which can be expressed in physical as well as organisational terms, and the demand for water, is not static and may vary considerably with time (Wilhite and Glantz, 1985). From a comparative study about drought policy in the United States and Australia, carried out by Wilhite (1986), it became clear that governments often respond to drought through crisis management rather than preplanned programmes (i.e., risk management). Hastily prepared assessment and response procedures may lead to ineffective, poorly coordinated, and untimely response (Wilhite and Wood, 1985; Wilhite, 1986).

Dams and reservoirs play a major role in a sustainable management of finite water resources; they have also become subject of vigorous public debates. It is argued that economic development and population growth, particularly in arid and semi-arid regions, make plain the need for dams for hydropower and irrigation (Altinbilek, 2002). The primary objective of water management is to determine the most effective operation strategy that maximizes the safety of water distribution and minimizes the threat of water shortage in time and space (Merabtene *et al.*, 2002).

Sedimentation reduces the storage capacity of reservoirs and thereby, their ability to conserve water for the various intended purposes. As a result the frequency and magnitude of their failure increases. Therefore, the effect of storage capacity losses on the water availability has to be considered within the management of available water resources, especially under water deficit conditions. At any dam or reservoir where sustainable long-term use should be achieved, it will be necessary to manage sediments as well as water (Morris and Fan, 1998).

1.2.2 Need for a risk management

Facing enormous economic and environmental costs to maintain and to restore water allocation and irrigation schemes, a sustainable water resources management in the Amu Draya basin is not so much realistic for the near future. It is also nearly impossible to develop strategies to compensate fully the impact of reduced water availability under current and future droughts. Instead, a first obligation must be to aspire in a development of risk management strategies a clear specification of (i) the risk anticipated (probability x loss), (ii) the losses and damage one is going to accept, and (iii) the options we have at hand to minimize the impact.

In general the evaluation and resolution of environmental and water related problems by a risk assessment and risk management techniques address the understanding of the extent and magnitude of hazards and the elaboration of practical and technically effective qualitative or quantitative goals and decisions regarding safe exposure levels, coupled with methods to minimize or control potential exposures (Teaf and Kuperberg, 2004).

Risk assessment basically provides a principled way of organising what we know about the world, particularly about its weak spots and creaky points (Jasanoff, 1993). Risk is defined as the probability of an outcome (e.g. a specified harmful effect will occur) multiplied by the severity of its consequence, leading to the potential quantification of risk in a number of ways. This notion can also be extended to positive as well as adverse impacts and the characterisation of uncertainty (Croke *et al.*, 2007).

Risk assessment is a methodological approach that provides a framework to acquire, organize and interpret data in order to provide the foundation for decisions regarding protection of human health and environmental quality. Approaches have been practised widely for about three decades in the context of contaminated sites, including the establishment of acceptable criteria for application to protection and rehabilitation of water resources, food supplies, air quality and the environment (Ganoulis, 2004, 2006; Croke *et al.*, 2007; Iglesias *et al.*, 2007). Paradigms for organising the elements of a risk-based approach have been articulated by such agencies as the World Health Organization (WHO, 2002), the U.S. Environmental Protection Agency (2002) and the European Commission (2002), among others.

The assessment of risk is not a goal in its own right but is part of an iterative cycle that uses the assessment of risk to derive management decisions that, when implemented, result in incremental improvements in water quality.

Risk assessment is a process complementary to impact assessment, with the latter used to assess the impacts of human activity. The environmental impact assessment, for example, is predictive, comparative and concerned with all possible effects on the environment.

The most important feature distinguishing risk assessment from impact assessment is the emphasis in risk assessment is on characterising and quantifying uncertainty. Thus, it is of particular interest in risk assessment to be able to analyse and estimate the analysable uncertainty.

Moreover, risk management is a decision-making process that applies the information from qualitative or quantitative risk assessments to establish a program that protects human health and the environment in the context of available technical and economic resources. Teaf and Kuperberg (2004) argued that risk management approaches are best applied in conjunction with a clear statement of potential risk and how they are being addressed.

Jain and Singh (2003) gave an example how risk management can be an integral part of the decision making process. They further described how the efficiency of risk assessment in water resources planning and management can be measured to provide assistance to planners and the decision makers.

1.3 ELEMENTS AND STEPS TO IMPLEMENT FUTURE RISK MANAGEMENT IN THE AMUDARYA BASIN

The central and most important project outcome is an analysis on future risk and the identification of sustainable management options for the water resources of the Aral Basin. There are many risks to the water resources but the main threats arises from climate change and a growing rural population, leading to less available water and an increased demand for irrigated land (increased food production) within the riparian states.

The condition of the water infrastructure and existing information systems are inadequate and do not allow an effective use of the abstracted water resource, resulting in a significant proportion of the command area being not irrigated or poorly irrigated in most years. Thus, there is excess wastage of water, a considerable amount of which has very high abstraction costs.

A regional water resources management strategy was developed by determining firstly the rate of glacial retreat and the consequences for future water resources as well as assessing of the likely impact of global changes on the annual precipitation in the region. A starting step of the investigations covers the estimation of the water consumption in the study area. Historic data were used and analysed applying water resources models. Those were first calibrated on an existing data basis and then used for simulating different scenarios up to 80 years into the future by combining the data gathered from the project.

From now to 2085 there will be growing demand for water to meet an evaporative demand arising from increased temperatures. Modelling indicates that in 2085 the evaporative demand is expected to increase between 7 and 19%, depending on the Global Circulation Model used and emission scenario adopted. With most outcomes being close to a 14% rise in evaporative demand, this equates to a mean increased water demand of about 3.5 km^3 for the area currently irrigated.

All outcomes from global models indicate little changes in precipitation in the upper catchment, but care is needed as those models hold a specific uncertainty simulating mountainous areas. Increased temperatures will result in shorter winters and a higher portion of the precipitation in the upper catchment will be rain rather than snow; an increased evapotranspiration and less runoff will be further the consequences. Glacier degradation is accelerating and by 2085 a notable reduction in glacial runoff is expected, which, although not contributing highly to the flow in wet years, makes a major contribution in drier years. Because of the difficulty in modelling future runoff in a catchment whose hydrological drivers are changing from those experienced in the past, accurate forecasts are not possible. Having explored a wide range of runoff scenarios, by 2085 at best we might expect to see a 5% reduction in flow and at worst a 20% reduction, which with increased evaporative demand equates to a mean shortfall between 10 and 25% of the existing water resource, but in dry years this could increase to 35%.

Unless irrigation practices change, with increasing evaporative losses and reduced river flow, by 2085 the area that can be irrigated will have to be reduced considerably. Model results indicate that, depending upon the used global circulation model and the CO_2 emission scenario, the decline in irrigable area during a quinquennial drought is likely to be in the rage of 40% of the existing command area; this is expected to be even higher in case the canal infrastructure will be not improved.

Although the scenario simulations show by trend that there will be less water available, and an increase in the evaporation, some uncertainty still exists, so no rushed decisions on management changes should be implemented. In contrast, well-considered and elaborated management strategies must be developed and seen in a wider scope considering mostly accompany options for improving the irrigation sector. It may not be possible to mitigate completely the adverse effects of climate change but there is significant potential to reduce them.

For example strategies for a drought-adapted water allocation need to be elaborated so that in periods of water shortage a sufficient supply under the prevailing circumstances can be also realised. As supplement for increasing available water for irrigation in dry years, non-saline renewable groundwater in the area of irrigation projects could be used. Groundwater provides inter-seasonal storage and can give a smoothing of the annual availability of the water resources. Furthermore, scavenger wells can be installed to tap fresh water lenses in an elsewise saline aquifer.

Within the irrigation massifs, where drainage does not return to the river during the irrigation season, water with salinity levels (Electrical Conductivity or EC) below 4 dSm, can be recycled directly into canals using low lift pumps for preventing wastage.

In addition, fallow fields in irrigation area should be avoided. Those lands attract lateral seepage from irrigated areas and thus create unwanted evapotranspirational losses from shallow groundwater, and by this decreasing the projects efficiency. In order to increase the productivity of the irrigated land the implementation of improved agronomic techniques are needed, such like laser grading.

Constraints on implementing strategies like described ones are the same as those that presently limit the performance of the irrigated sector, namely the lack of investment and the pressuring need to modernise irrigated agriculture. Delaying investments in all sectors of water management and irrigated agriculture will only accelerate the decline in areas that can be irrigated, and thus increase its severity.

For the first time, the project JAYHUN teams sophisticated estimations on future water availability with an interstate reservoir management to address further pressuring water scarcity in the Aral Sea Basin. The results are high innovative risk management strategies, which clearly progress the state of the art beyond previous fragmented approaches. Secondly, the involvement of Tajik partners and key institutions controlling the primary dams in the upstream region has broken new ground with respect to an interstate dialogue and the development of more realistic water allocation patterns since the collapse of the Soviet Union.

The JAYHUN project further managed bringing together a range of modern tools for water resource planning that provide solutions for complex systems in a large basin. The project's real innovation lies not in each single tool but in their combination provided as an integrated application package. In the past, considerable resources have been spent on investigating the Aral Sea Crisis but unfortunately most studies had not addressed the issue of the reliability of basic data. In the course of the JAYHUN project, cost effective tools were implemented which are capable of acquiring data efficiently over the vast area of the basin. This approach was essential since the data collection system had largely collapsed, and analyses show that a considerable amount of existing official data is of insufficient quality and certain uncertainty.

Given the still growing pressures on the water allocation in the region, it is likely that in the absence of reliable information disputes will increase with resulting impacts at regional, national, and even international level. The approach developed within JAYHUN addresses these challenges and makes a significant contribution to providing reliable estimates of the water availability and thus assists resolving regional tensions and conflicts on water issues. Moreover, the riparian states gained essential information for an effective water resource planning and management at basin level.

In addition to deepen the collaboration between European and Central Asian research institutions, JAYHUN had also been successful in providing an expandable platform and forum for scientific and institutional exchanges allowing a wider understanding of the problems in the Amu Darya Region. Furthermore it established authentic boundary conditions to achieve a sustainable water resources management in the Aral Sea Basin. Without those conditions, the needed level of sustainability will most certainly not be possible.

1.4 REFERENCES

ADB (Asian Development Bank) (2004) Program concept for pipeline entry for the Global Environment Facility for Central Asian Countries Initiative for Land Management, at: http://www.gefonline.org/projectDetails.cfm?projID = 2504, 29 March 2006.

Ahmad, M. and Wasiq, M. (2004) Water Resource Development in Northern Afghanistan and its implication for Amu Darya Basin, World Bank Working Paper No. 36.

Altinbilek, D. (2002) The role of dams in development. *Water Science and Technology*, **45**(8), 169–180.

Croke, B.F.W., Ticehurst, J.L. *et al.* (2007) Integrated assessment of water resources: Australian experiences. *Water Resources Management*, **21**(1), 351–373.

European Commission (2002) European Commission Chemical Risks, website: http://ihcp.jrc.cec.int/Activities/ACTChem/ACTChem.html

Froebrich, J. and Kayumov, O. (2004) Water management aspects of Amu Darya. In *Dying and Dead Seas – Climatic Versus Anthropic Causes.* Nihoul *et al.* (eds.), Nato Science Series: IV, Earth and Environmental Sciences. Kluwer Academic Publishers, Dordrecht, Vol. 36, pp. 49–76.

Ganoulis, J. (2006) Water resources management and environmental security in mediterranean transboundary river basins. In *Environmental Security and Environmental Management: The Role of Risk Assessment.* Morel, B. and Linkov, I. (eds.), Springer, pp. 49–58.

Ganoulis, J. (2004) Integrated risk analysis for sustainable water resources management. In *Comparative Risk Assessment and Environmental Decision Making.* Linkov, I. and Bakr Ramadan, A. (eds.). NATO Science Series, Kluwer Academic, The Netherlands, pp. 275–285

Giese, E. and Mossig, I. (2004) Klimawandel in Zentralasien, Zentrum internationale Entwicklungs- und Umweltforschung der Justus – Liebig-Universität Gießen, Discussion Papers, Nr.17.

ICWC (2008) Interstate Commission for Water Coordination of Central Asia, http://www.icwc-aral.uz/index.htm

Iglesias, A., Garrote, L., Flores, F. and Moneo, M. (2007) Challenges to manage the risk of water scarcity and climate change in the Mediterranean. *Water Resources Management*, **21**(5), 775–788.

IPCC (2001) Climate Change 2001: The Scientific Basisi. Contribution of Working Group I to the Third Assessment Report of the Intergovernmental Panel on Climate Change. Cambridge University Press, IPCC, pp. 1–881.

Jain, S.K. and Singh, V.P. (2003) Water Resources Systems Planning and Management. Developments in Water Science, Vol. 51. Amsterdam, Elsevier.

Jasanoff, S. (1993) Bridging the two cultures of risk analysis. *Risk Anal.*, **13**, 123–129.

Lange, K. (2001) Energy and Environmental Security in Central Asia: The Syr Darya. Centre for Strategic and International Studies (CSIS), http://www.csis.org/ruseura/cs010220lange.htm, Accessed May, 2006.

Loucks, D.P. and Gladwell, J.S. (eds.) (2000) Sustainability criteria for water resource systems. Prepared by the working group of UNESCO/IHP Project M-4.3. Cambridge University Press.

Martius, C., Froebrich, J. and Nuppenau, E.A. (2007) Conceptionalizing integrated, transnational water resource management in the Amu Darya watershed. In *Facing Global Environmental Change Vol. 2 , Environmental, Human, Energy, Food, Health and Water Security Concept, Hexagon Series on Human and Environmental Security and Peace.* Brauch, H.G., Czeslaw Mesjasz, J.G., Chadha Behera, N., Chourou, B., Oswald Spring, U., Liotta, P. H., Kameri-Mbote, P. (eds.), Springer Verlag, Berlin – Heidelberg – New York – Hong Kong – London – Milan – Paris – Tokyo.

MEA (Millenium Ecosystem Assessment) (2005) Ecosystems and Human Well-being: Desertification Synthesis. Washington, DC: Word Resources Institute.

Merabtene, T., Kawamura, A., Jinno, K. and Olsson, J. (2002) Risk assessment for optimal drought management of an integrated water resources system using a genetic algorithm. *Hydrological Processes*, **16**(11), 2189–2208.

Morris, G. and Fan, J. (1998) Reservoir Sedimentation Handbook. McGraw-Hill, New York.

National Commission of the Republic of Uzbekistan on Climate Change (1999) Initial Communication of the Republic of Uzbekistan under the United Nations Framework Convention on Climate Change.

O'Hara, S. (2000) Central Asia's water resources: Contemporary and future management issues. *Water Resource Development*, **16**(3), 423–441.

Olsson, O., Ikramova, M., Bauer, M. and Froebrich, J. (2009) Applicability of Adapted Reservoir Operation for Water Stress Mitigation Under Dry Year Conditions. Water Resources Management, DOI 10.1007/s11269-009-9446-x (available online).

Smith, D.R. (1995) Environmental security and shared water resources in post-Soviet Central Asia. *Post Soviet Geography*, **36**(6), 351–370.

Tanton, T. and Heaven, S. (1999) Worsening of the Aral Basin Crisis can there be a solution? *Journal of Water Resources Planning and Management*, **125**(6), 363–368.

Teaf, C.M. and Kuperberg, J.M. (2004) Risk assessment, risk management & risk communication: principles and applications. In *Risk Assessment as a Tool for Water Resources Decision-Making in Central Asia.* Teaf *et al.* (ed.). NATO Science Series, Kluwer Academic, The Netherlands, pp. 1–15.

UNFCCC (Initial National Communication of the Republic of Uzbekistan under the UN Framework Convention on Climate Change) (2001) Uzbekistan: "Country study on climate change, phase 2. Expedited financing of climate change enabling activities. Project implementation report. Main administration under the Cabinet of Ministers; Uzbekistan (Mimeo).

U.S.EPA (2002) U.S. Environmental Protection Agency Superfund Risk Assessment, website: http://www.epa.gov/superfund/programs/risk/index.htm

Wegerich, K. (2004) Coping with disintegration of a river-basin management system: multi-dimensional issues in Central Asia. *Water Policy*, **6**, 335–344.

Wegerich, K. (2009) The hidden urban tail-enders – drinking water supply as a common pool resource problem in Khorezm, Uzbekistan. *Int. J. Sustainable Society*, **1**(3), 240–254.

Wegerich, K., Olsson, O. and Froebrich, J. (2007) Reliving the past in a changed environment: Hydropower ambitions, opportunities and constraints in Tajikistan. *Energy Policy*, **35**, 3815–3825.

WHO (2002) World Health Organization IPCS Risk Assessment, website: http://www.who.int/pcs/ra_main.html

Wilhite, D.A. (1986) Drought policy in the U.S. and Australia: A comparative analysis. *Water Resources Bulletin*, **22**, 425–438.

Wilhite, D.A. and Glantz, M.H. (1985) Understanding the drought phenomenon: The role of definitions. *Water International*, **10**, 111–120.

Wilhite, D.A. and Wood, D.A. (1985) Planning for drought: The role of state government. *Water Resources Bulletin*, **21**, 31–38.

Chapter 2

Modelling the future climate of the Amu Darya Basin

Natalya Agaltseva, Tatjana Spectorman, Christopher White and Trevor Tanton

2.1 IPCC 4TH CLIMATE CHANGE ASSESSMENT

Climate change has introduced a man made risk affecting the future reliability of the water resources in the Amu Darya Basin. The magnitude and variability of this change is poorly understood and, until recently, has been largely unquantifiable. However, the International Panel on Climate Change (IPCC) has collaboratively with numerous climate institutions around the world provided a suite of Global Circulation Models (GCMs) in the IPCC 4th Assessment Report (AR4, 2007) which allow for the quantification and assessment of the likely changes to the climate under various greenhouse gas emission scenarios.

The range of emission scenarios developed by the IPCC describes developments in many different social, economic, technological, environmental and policy dimensions (Figure 2.1). We do not, and never will, know how these will change in the future. The best we can do is to envisage several plausible ways in which the world will develop over the next century.

The IPCC clearly states that it is not possible to put relative probabilities on any of the emissions scenarios (i.e. they may not be equally probable, nor should any of the scenario groups be discounted). It is noticeable however that, despite the immediate divergence of future emissions in the scenarios, the warming over the next 40 years from each emissions scenario is very similar (Figure 2.2). This is mainly due to the very long effective lifetimes of CO_2 and the large inertia of the climate system. Much of the warming over the next few decades is already built into the climate system due to current emissions and those over the past several decades.

The medium/high emission scenario (A2) describes a very heterogeneous world. The underlying theme is self-reliance and preservation of local identities. Fertility patterns across regions converge very slowly, which results in high population growth. Economic development is primarily regionally oriented and per capita economic growth and technological change are more fragmented and slower than in other scenarios. The authors consider that, given current knowledge, it is probably the most realistic view *at this time*. Nevertheless, it is only one possible scenario, but it is important to note that all other possible future scenarios developed by the IPCC predict significant increases in temperature in the Amu Darya Basin that are well above the global average.

Projections from the various GCMs do not have an equal probability of representing the true outcome, but evaluating the relative credibility of each model (and hence their projections) is not straightforward to do, and requires a detailed knowledge of climate modelling and the climate system, together with a comprehensive database of observations.

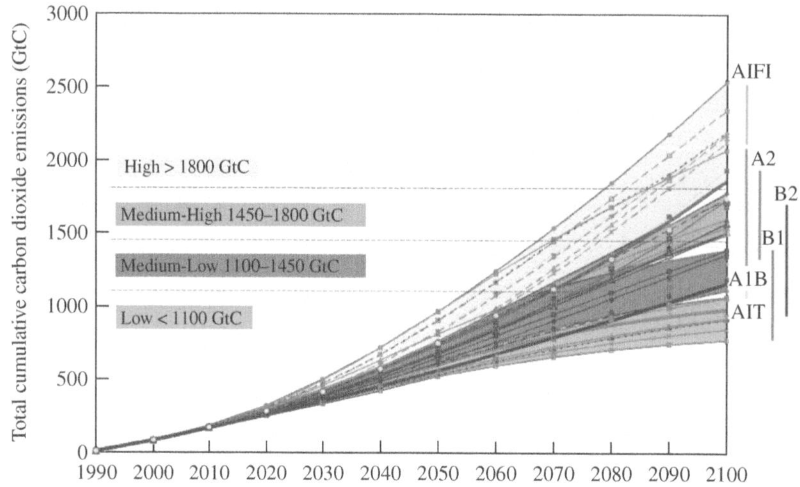

Figure 2.1. IPCC global cumulative carbon dioxide emissions scenarios from 1990 to 2100 for each scenario, grouped by family and group (Nakicenovic *et al.*, 2001).

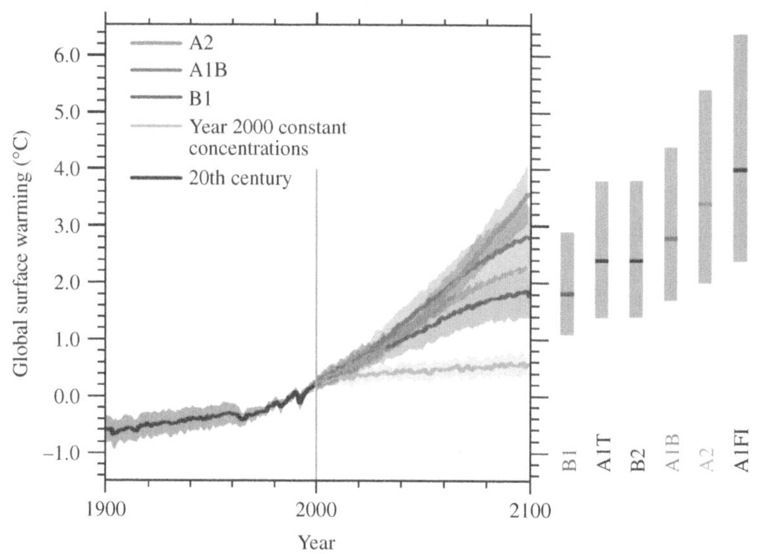

Figure 2.2. Multi-model averages and assessed ranges for surface warming. Solid lines are multi-model global averages of surface warming (relative to 1980–1999) for the scenarios A2, A1B and B1, shown as continuations of the 20th century simulations. Shading denotes the plus/minus one standard deviation range of individual model annual averages (IPCC 4th Assessment Report).

The model outputs are however at a relatively coarse global scale. Planners and risk assessors require data at a more regional or local scale to determine the potential impact of climate change on water resources over the coming century. The Amu Darya Basin is divided into two main zones – a mountainous zone in the east and a plains zone in the west. The mountains, up to 7000m form part of the Pamir, Hindu Kush, and Tien Shan ranges. The plains form a series of three major terraces stepping down to the margins of the Aral Sea.

In this study of the Amu Darya Basin, two spatial downscaling methodologies were applied for regional climate scenario construction. For the planes part of Amu Darya Basin downscaling methodology was based on an interpolated high-resolution observed dataset which was applied to the modelled projections extracted from the GCMs to provide likely future climatologies at specific locations of interest throughout the region. For the mountains part of Amu Darya Basin the statistical interpretation method was applied. This method is based on multiple linear regression, equations between data in grid nodes and observation stations were constructed but applied to the GCMs output data from MAGICC/SCENGEN software (Wigley, 2003).

This study provides a suite of downscaled future climatic projections under the various emissions scenario to assess the potential long-term impact of climate change on the water resources of the Amu Darya Basin. The output is designed to allow strategic studies to identify potential future constraints to the water supply and demand, changes in river flows and runoff, and to estimate potential impact on the future use and development of irrigation and drainage systems in the region.

2.1.1 Regional Climate, Amu Darya Basin

The Amu Darya Basin is part of the larger Aral Sea Basin, which extends principally to the east and south of the Aral Sea. In total, the Aral Sea Basin covers an area of about 1.5 million km^2 of which about 0.6 million km^2 is classed as cultivable. The basin contains the catchments of not only the Amu Darya River, but also the Syr Darya River which both flow westward from the mountain ranges to the Aral Sea.

The topography of the Amu Darya Basin is divided into two main zones – a mountainous zone in the east and a plains zone in the west. The mountains, up to 7000 m form part of the Pamir, Hindu Kush, and Tien Shan ranges. The plains form a series of three major terraces stepping down to the margins of the Aral Sea.

Due to the Amu Darya Basin's central location in the Asian landmass, the climate in the region is strongly continental in nature, characterised by generally low precipitation, low relative humidity, cold winters and hot summers. The annual precipitation in many of the lowlands and valleys is typically very low, ranging from less than 100 mm to 200 mm, concentrated mainly in the winter and spring. Virtually no rainfall occurs during the summer months. Precipitation increases towards the mountains, reaching up to 800 mm per year along the southern and south-western sides of the ranges and over 1000 mm on the highest peaks. Figure 2.3 shows the temporal and spatial temperature and precipitation change over the last century for the greater Aral Sea Basin including the Amu Darya Basin.

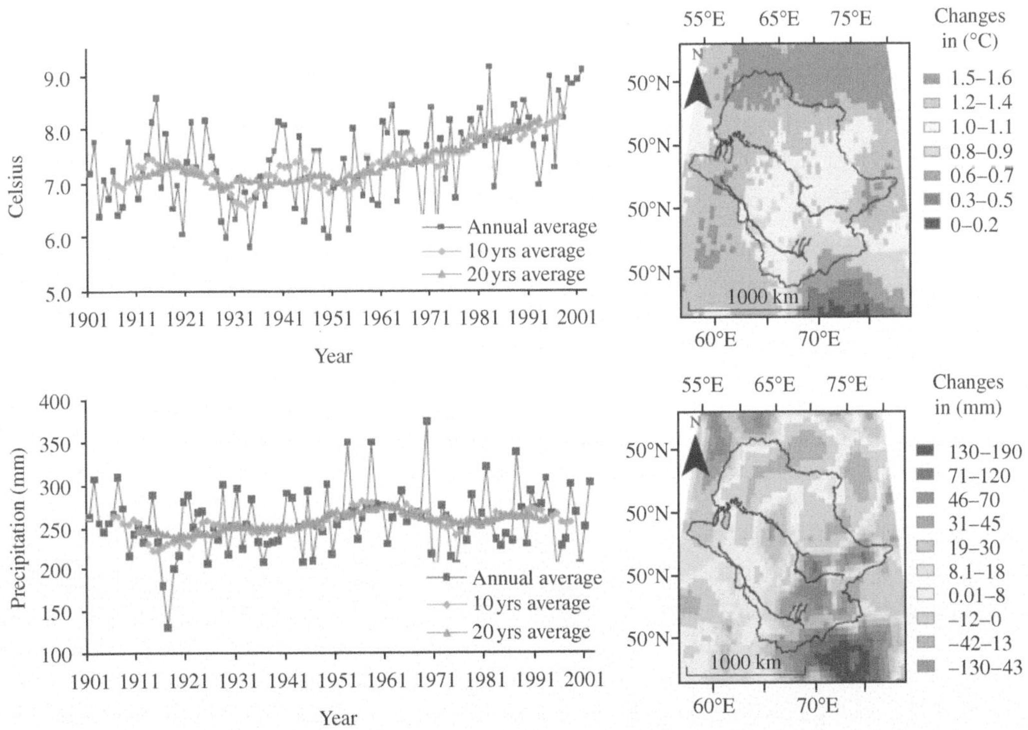

Figure 2.3. Temperature (top) and precipitation (bottom) data within the Aral Sea Basin [Note: Temporal trends (left) and spatial distribution of change (right) between the temporal (50 years) average pre-1950 condition and the recent 20 years (1983–2002) average condition (Shibuo *et al.*, 2007).

2.2 CLIMATE PROJECTIONS FOR THE PLAINS PART OF AMU DARYA BASIN

2.2.1 Global Circulation Models (GCMs)

For this research, seven GCMs were selected to determine a range of possible future climate projections for the Amu Darya region (Table 2.1) to allow for a range of possible outcomes for the same scenario.

Table 2.1. IPCC AR4 GCMs used to extract climate projections for the Amu Darya Basin.

GCM Model ID	Organisation	Country	Grid Resolution
CCSM3 (2005)	(NCAR)	USA	1.4°N×1.4°E
CNRM-CM3 (2004)	(CNRM)	France	~2.8°N×2.8°E
CSIRO-MK3.0 (2001)	Commonwealth Scientific & Industrial Research Organisation (CSIRO)	Australia	~1.9°N×1.9°E
ECHAM5 (2005)	Max Planck Institute for Meteorology (MPIM)	Germany	~1.9°N×1.9°E
GFDL-CM2.0 (2005)	Geophysical Fluid Dynamics Laboratory (GFDL/NOAA)	USA	2.0°N×2.5°E
HadCM3 (1997)	British Meteorological Service Hadley (UKMO)	UK	2.5°N×3.5°E
MIROC3.2(MED) (2004)	Centre for Climate System Research (CCSR), University of Tokyo	Japan	~2.8°N×2.8°E

Source: IPCC-DCC website.

The GCMs were obtained with varying grid resolutions specific to each of the climate institutions around the world, ranging from 1.4°N×1.4°E (CCSM3) to 2.5°N×3.5°E (HadCM3). Summary data (30-year means) was provided in netCDF format (.nc files) for the IPCC AR4 global means and GRIB format (.grb files) for the IPCC AR4 monthly means. These files were converted to ASCII files using the ncdump.exe and grbconv.exe MS-DOS programs which are freely available from the IPCC-DCC website.

The averaged time periods considered were 1961–1990 for the hindcast reanalysis (CRU observations and 20C3M experiment), and 2070–2099 for the GCMs future climate projections (A2 SRES scenario) of monthly and 30-year means. [Note: For presentation purposes in this report, only the historical 1961–1990 period and the future 2070–2099 projections are used, however data for 1901–1930, 1931–1960 (hindcast), 2010–2039 and 2040–2069 (future projections) is available from the website address: http://www.inco-jayhun.net]

2.2.2 Assessment of Climate Model Performance (1961–1990)

Methodology

Modelled temperature and precipitation data were extracted from a selection of seven GCMs [Table 2.1; source: IPCC-DCC (Data Distribution Centre) website] for the Amu Darya Basin for the 20C3M experiment. The 20C3M experiment provides a modelled hindcast dataset for the 20th Century, forced using observed data, to allow for correlation and reanalysis of the modelled output with historical data. This provides modellers with the necessary information with which to assess the relative merits of each GCM (and ensembles of GCMs) for any given location before using or applying future climate projections for selected scenarios.

The individual performance of each of the seven GCM 20C3M experiments was reanalysed using the observed Climate Research Unit (CRU) high-resolution global climate dataset for the last century, covering the averaged time period of 1961–1990. This was undertaken at eleven selected locations across the Amu Darya Basin.

CRU high-resolution datasets

The CRU high-resolution global climate dataset consists of a multi-variate 0.5° latitude by 0.5° longitude resolution monthly mean climatology for all global land areas. This was obtained for the Amu Darya Basin region together with a 30-year mean climatology, a monthly time series at the same resolution for the period 1901–2002 was also obtained from the Tyndall Centre, UK titled as the CRU TS 2.1 monthly means dataset.

The mean climatologies are a suite of surface variables constrained to the averaged time period of 1961–1990 to provide an observed baseline normal. Those used in this study were precipitation (PR) and 2m surface mean temperature (TMP). The mean climate surfaces have been constructed from a dataset of station climatological normals. The station data were interpolated as a function of latitude, longitude and elevation using thin-plate splines. The accuracy of the interpolations were assessed using cross-validation and by comparison with other climatologies (New *et al.*, 1999).

It was noted by the authors that the CRU TS 2.1 is not specifically designed for the actual detection of anthropogenic climate change (e.g. Patz *et al.*, 2002). This issue concerns the distinctly anthropogenic climate changes that may already be being experienced in the Amu Darya Basin. In contrast however, the primary purpose for which CRU TS 2.1 was constructed was to permit modellers to incorporate an accurate a representation as possible of month-to-month climate variability as experienced in the recent past as demonstrated by Shibuo *et al.* (2007). As the aim here was not to detect climate change from the CRU dataset, but to simply use it as a baseline period from which to add or subtract the climate projection anomalies from the GCMs, the use of the CRU datasets was deemed to be appropriate for this study. However, the authors do note that the CRU TS 2.1 dataset does not necessarily give a homogenous representation of climate change for every individual grid cell due to the interpolation. This is not a fault, but an inevitable consequence of the design, especially in a region like the Amu Darya Basin where observed data is particularly sparse and known to have reliability issues. The CRU TS 2.1 provides a 'best estimate' of the spatial pattern of climate at each moment in time - it is complete in space and time. Because the emphasis is placed on obtaining the most accurate estimates of the instantaneous spatial patterns, the authors note that in homogeneities may be present in the time series of individual grid cells.

Selected locations in the Amu Darya Basin

The reanalysis was carried out using eleven specific locations of interest across the Amu Darya Basin (Table 2.2). These locations were selected to satisfy a range of interests (irrigation schemes, urban centres and high altitude sites).

Table 2.2. Eleven locations across the Amu Darya Basin for the reanalysis of observed CRU high-resolution data & GCM 20C3M modelled data (1961–1990).

Location	Coordinates (relative to each CRU Global Grid)	Elevation (at the centre of the CRU Global Grid)
Bukhara	39.75°N, 64.25°E	234 m
Dashauz	41.25°N, 59.75°E	99 m
Down Kafirnigan	37.25°N, 67.75°E	305 m
Dushanbe	38.25°N, 69.25°E	1420 m
Karakalpakstan	42.25°N, 59.25°E	67 m
Lebab	39.25°N, 63.25°E	181 m
Mary/Ahal	37.25°N, 61.75°E	244 m
Nukus	42.75°N, 59.25°E	70 m
Surxhondaryo	37.75°N, 67.75°E	475 m
Urgench	41.25°N, 60.25°E	104 m
Vaksh	37.25°N, 68.75°E	330 m

Reanalysis of observed CRU high-resolution data & GCM 20C3M modelled data for the Amu Darya Basin

Table 2.3 shows annual data for temperature and precipitation variation for the modelled data against the CRU high-resolution database for the historical period of 1961–1990.

Analysis of the data shows similarities and correlation between outputs from the modelled 20C3M data and the observed CRU data at some, but not all, of the eleven locations in the Amu Darya Basin. For temperature, locations such as Bukhara and Lebab show a close correlation between observed and modelled data for the 1961–1990 period. However, other locations such as Down Kafirnigan, Dushanbe and Surxhondaryo display a poor relationship (Figures 2.4–2.7).

Further interrogation of the data revealed that, for the poorly correlated locations with high variability and range of values across the GCMs, the relative performance of the models was not uniform (i.e. some GCMs performed notably worse than others). This was found to be inconsistent however, with some models performing poorly at one location, but better at others. The ECHAM5 model is such an example, where it performed well when modelling temperatures at Surxhondaryo, against the trend of the other models that all perform particularly poorly. However, the ECHAM5 model then produced poor results for precipitation at several of the eleven locations. This was a similar pattern repeated across the model outputs and locations. The HadCM3 model matched temperature quite well across all sites but as with most models was inconsistent in modelling precipitation.

Several conclusions may be drawn. Both the CRU observations and the 20C3M data have inherent differences in the way they are produced – neither of which may produce completely accurate results. When compared, these differences may be compounded. Whilst the CRU data is based on observations, it still remains an interpolated dataset, which is open to errors, in homogeneities and altitude inconsistencies that may effect the correlation with the modelled 20C3M data. However, it was concluded that the variability between the model outputs and the large grid sizes of the GCMs compared to the high-resolution of the CRU is more likely to have caused the large inconsistencies between the observed and modelled data together with the differing approaches of all the GCMs. For example, where a CRU grid cell falls on the edge of one of the GCMs, the modelled data may not be as accurate were the CRU grid cell to be nearer to its centre. In some cases, up to twelve CRU grid cells occupy a single GCM, leading to the conclusion that the large GCM resolutions are a significant source of error in this analysis.

Table 2.3. AR4 20C3M 2 m annual mean surface temperature (TMP) and precipitation (PR) climatologies (1961–1990) compared to the observed CRU 1961–1990 baseline annual mean normals for the eleven locations across the Amu Darya Basin.

Locations across the Amu Darya Basin	GCMs (20C3M), temperature (°C)								GCM (20C3M) Ensemble	
	CRU	**CCSM3**	**CNRM-CM3**	**CSIRO-MK3.0**	**ECHAM5**	**GFDL-CM2**	**HadCM3**	**MIROC3.2**	**Mean TMP**	**TMP Range**
Bukhara	15.44	14.63	14.62	14.47	15.90	14.51	14.32	16.37	14.97	2.05
Dashauz	13.28	14.13	16.98	12.53	15.36	14.20	12.52	18.50	14.89	5.97
Down Kafirnigan	17.13	9.07	11.18	4.07	11.53	5.37	12.98	10.98	9.31	8.91
Dushanbe	13.58	7.42	2.34	2.62	9.55	3.56	12.98	5.10	6.22	10.64
Karakalpakstan	12.06	13.35	14.66	12.53	15.36	12.32	12.52	17.57	14.04	5.25
Lebab	15.69	15.69	16.15	14.47	15.90	14.51	14.32	17.92	15.57	3.61
Mary/Ahal	16.72	14.03	16.27	11.44	16.18	14.35	14.61	18.37	15.04	6.93
Nukus	11.48	13.35	14.66	12.53	15.36	12.32	12.52	17.57	14.04	5.25
Surxhondaryo	15.68	9.07	11.18	9.87	15.64	5.37	12.98	10.98	10.73	10.27
Urgench	13.39	13.94	16.98	12.53	15.36	14.04	12.52	18.50	14.84	5.97
Vaksh	17.23	5.55	2.34	2.62	9.55	5.37	12.98	5.10	6.21	10.64
	GCMs (20C3M), precipitation (mm)								**Mean PR**	**PR Range**
Bukhara	157.06	176.96	102.09	82.56	120.26	232.55	100.04	213.26	146.82	149.99
Dashauz	109.58	190.87	82.17	117.24	74.53	149.99	100.04	145.05	122.84	116.35
Down Kafirnigan	266.63	354.98	264.01	326.49	316.06	742.77	422.75	451.33	411.20	478.76
Dushanbe	493.09	459.41	704.95	440.77	640.32	853.86	422.75	629.85	593.13	431.11
Karakalpakstan	109.58	226.29	78.12	117.24	74.53	166.71	100.04	161.95	132.13	151.77
Lebab	138.80	153.82	60.74	82.56	120.26	232.55	100.04	147.11	128.15	171.81
Mary/Ahal	171.67	182.14	227.15	146.66	141.20	244.34	149.94	174.46	180.84	103.15
Nukus	116.88	226.29	78.12	117.24	74.53	166.71	100.04	161.95	132.13	151.77
Surxhondaryo	394.47	354.98	264.01	249.65	365.10	742.77	422.75	451.33	407.23	493.13
Urgench	109.58	183.92	82.17	117.24	74.53	138.31	100.04	145.05	120.18	109.40
Vaksh	328.73	436.21	704.95	440.77	640.32	742.77	422.75	629.85	573.95	320.02

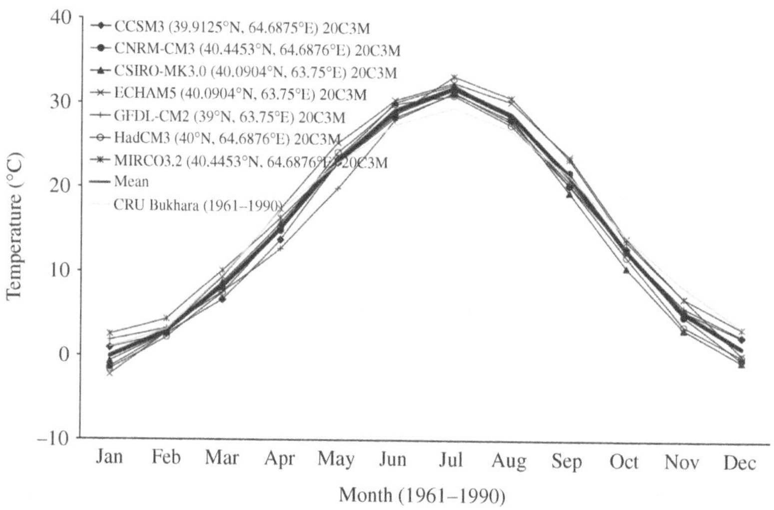

Figure 2.4. AR4 20C3M 30-year 2 m mean surface temperature climatologies (Jan–Dec, 1961–1990) compared to the observed CRU 1961–1990 baseline normal for Bukhara.

Ensemble means of the seven GCMs were also calculated and correlated again the CRU data for 1961–1990 for both temperature and precipitation. Table 2.4 shows annual data for temperature (°C) and precipitation correlations for the modelled data against the CRU high-resolution database for the historical period of 1961–1990.

Whilst the ensemble values modelled the CRU observations better than *most* of the individual GCMs, the results of the GCM ensemble means were not always statistically significant when compared to the CRU high-resolution data,

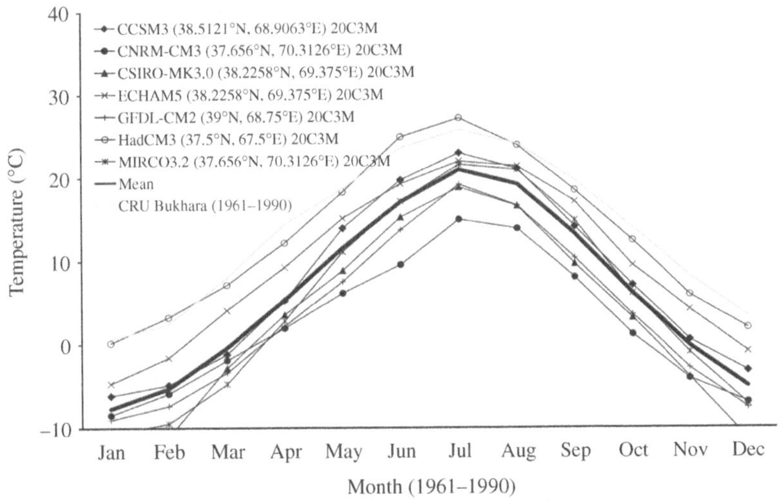

Figure 2.5. AR4 20C3M 30-year 2 m mean surface temperature climatologies (Jan–Dec, 1961–1990) compared to the observed CRU 1961–1990 baseline normal for Dushanbe.

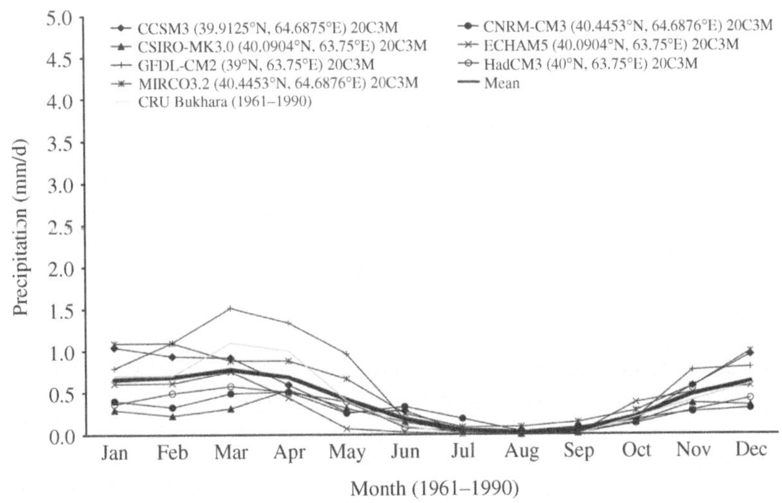

Figure 2.6. AR4 20C3M 30-year precipitation climatologies (Jan–Dec, 1961–1990) compared to the observed CRU 1961–1990 baseline normal for Bukhara (39.75° N, 64.25° E).

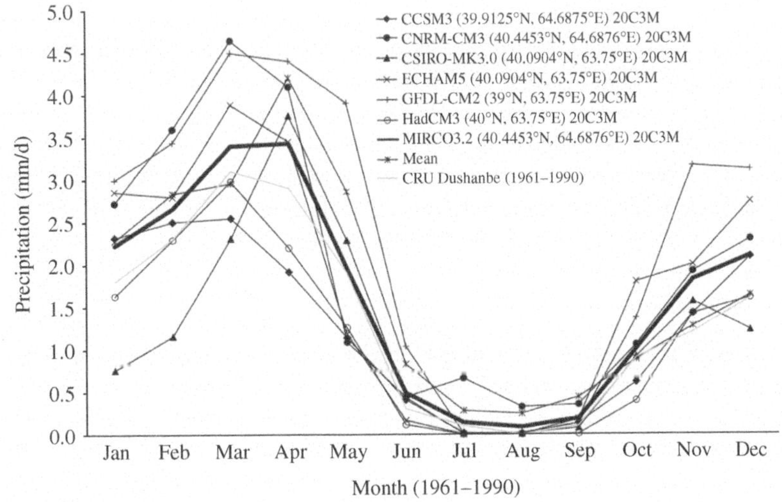

Figure 2.7. AR4 20C3M 30-year precipitation climatologies (Jan–Dec, 1961–1990) compared to the observed CRU 1961–1990 baseline normal for Dushanbe.

Table 2.4. AR4 20C3M 30-year 2 m mean surface temperature and precipitation correlations for the modelled data against the CRU high-resolution database for the historical period of 1961–1990 for the eleven locations across the Amu Darya Basin (R^2 and P-values calculated using monthly means).

Locations across the Amu Darya Basin	AR4 GCM (20C3M) Ensemble (1961–1990)			
	TMP		PR	
	R^2	P-value	R^2	P-value
Bukhara	0.9902	0.1883	0.9238	0.2252
Dashauz	0.9889	0.0005	0.7863	0.1208
Down Kafirnigan	0.9915	0.0000	0.9261	0.0009
Dushanbe	0.9927	0.0000	0.9805	0.0004
Karakalpakstan	0.9886	0.0002	0.6285	0.0721
Lebab	0.9894	0.4176	0.9626	0.2373
Mary/Ahal	0.9909	0.0002	0.8980	0.2446
Nukus	0.9873	0.0000	0.6860	0.1159
Surxhondaryo	0.9924	0.0000	0.9755	0.2413
Urgench	0.9891	0.0012	0.7974	0.1646
Vaksh	0.9851	0.0000	0.9259	0.0005

with one (or more) of the individual GCMs typically displaying a higher statistical significance compared to the CRU high-resolution data than the ensemble means. This was thought to be due to the loss of the individual GCM model integrity when averaging the results. For example, one GCM may perform better for the summer months than winter, but this information may get lost or may skew the ensemble means. The temperature variability was however generally found to be, in many cases, relatively low while precipitation variability was significant when compared to the CRU 1961–1990 baseline observations.

The authors conclude that, whilst adequate for the assessment of performance of individual GCMs against observed data, the 20C3M data pertaining to each GCM was not of sufficient quality (in range, variability or magnitude) to be used as a 'historical' baseline period from which to relate future climate projections for the Amu Darya Basin. The 20C3M experiment data produced good trends together with good monthly and seasonal variability, but for absolute climatologies (as shown here) the overall performance produced inconsistent results. The high degree of disagreement between the models, the coarse grid resolutions, the lack of adequate local time series climate observations and the crudeness of the elevations inherent in the GCM climatologies determined that the CRU high-resolution dataset would be used in-lieu of the 20C3M or local observed data to downscale the future climate projections to the local scale for the Amu Darya Basin.

Downscaled future climate projections

It was decided that the 20C3M hindcast experiment data from the suite of GCMs would not provide an adequate baseline 1961–1990 period from which to relate the future anomaly projections. Instead, procedures for downscaling the GCM model outputs from the coarse global resolutions to high-resolution grid cells were examined, using such factors as topography, single observed time series and interpolated regional or global temperature and precipitation datasets.

As the resolution of the seven GCMs used in this study were of typically low (and differing) grid resolutions, and taking into account the requirements for higher resolution by the applied water balance models, a spatial downscaling procedure was considered necessary and appropriate. The interpolated CRU high-resolution dataset (the CRU 30-year mean and the CRU TS 2.1 monthly means, January to December) were applied to the GCM output values to provide a downscaled baseline for the 1961–1990 climatology. Future climate projection anomalies from the GCMs for the averaged time period of 2070–2099 were then extracted and spatially downscaled for specific locations of interest throughout the Amu Darya Region using the high-resolution (0.5° latitude by 0.5° longitude) CRU global observed data.

The key reference period of 2070–2099 has been adopted in this study to demonstrate the future risk to the water resources of the Amu Darya Basin. However, downscaled future climate projections for 2010–2039 and 2040–2069 can also be viewed on the website address: http://www.inco-jayhun.net

Temperature: The A2 scenario represents a future with a high global increase in average temperature anomalies from the baseline 1961–1990 means. Under the A2 scenario, future climate variability is also anticipated in the Amu Darya Basin. In this study, all of the GCMs used indicate that it is likely that it will get appreciably hotter in the Amu Darya Basin by the period 2070–2099 when compared to the 1961–1990 averaged annual baseline data (4.35°C on average across the eleven locations). For the time period of 2070–2099 (Tables 2.5 and 2.6), the changes in annual means demonstrate significant consistency between the GCMs at all of the eleven locations, with between a 25% and 38% increase in mean temperature projected by 2099 across the region.

Table 2.5. A2 scenario (2070–2099) 2 m annual mean surface temperature anomalies (°C) relative to the observed CRU 1961–1990 baseline annual mean normals for the eleven locations across the Amu Darya Basin.

Locations across the Amu Darya Basin	AR4 GCMs (A2 Scenario, 2070–2099)								
	CCSM3	CNRM-CM3	CSIRO-MK3.0	ECHAM5	GFDL-CM2	HadCM3	MIROC3.2	Mean TMP Anomaly	Range
Bukhara	4.75	4.69	3.58	4.75	3.83	4.42	4.48	4.36	1.17
Dashauz	4.72	4.69	3.55	4.63	3.66	4.82	4.41	4.35	1.28
Down Kafirnigan	4.95	4.82	3.80	5.13	4.98	4.43	5.00	4.73	1.33
Dushanbe	5.19	4.96	4.05	5.24	4.67	4.43	5.42	4.85	1.37
Karakalpakstan	4.65	4.78	3.55	4.63	3.69	4.82	4.77	4.41	1.28
Lebab	4.79	4.68	3.58	4.75	3.83	4.42	4.43	4.36	1.21
Mary/Ahal	4.77	4.54	3.45	4.56	3.90	4.45	4.43	4.30	1.31
Nukus	4.65	4.78	3.55	4.63	3.69	4.82	4.77	4.41	1.28
Surxhondaryo	4.95	4.82	3.74	4.83	4.98	4.43	5.00	4.68	1.26
Urgench	4.76	4.69	3.55	4.63	3.67	4.82	4.41	4.36	1.28
Vaksh	5.45	4.96	4.05	5.24	4.98	4.43	5.42	4.93	1.40

Table 2.6. A2 scenario ensemble (2070–2099) of 2 m annual mean surface temperature climatologies (°C) downscaled using the observed CRU 1961–1990 baseline annual mean normals for the eleven locations across the Amu Darya Basin. (Standard deviations, confidence intervals and percentage change estimates calculated using monthly means.)

Locations across the Amu Darya Basin	CRU (1961–1990)	AR4 GCM Ensemble (A2 Scenario, 2070–2099)						
		Mean TMP Anomaly	Mean TMP	Standard Deviation	95% Conf. Interval (Lower)	95% Conf. Interval (Upper)	95% Conf. Level	% Change
Bukhara	15.44	4.36	19.80	0.47	19.45	20.15	0.70	28.23
Dashauz	13.28	4.35	17.63	0.53	17.24	18.02	0.79	32.79
Down Kafirnigan	17.13	4.73	21.86	0.47	21.52	22.21	0.69	27.60
Dushanbe	13.58	4.85	18.43	0.49	18.07	18.80	0.73	35.72
Karakalpakstan	12.06	4.41	16.47	0.55	16.07	16.88	0.81	36.59
Lebab	15.69	4.36	20.05	0.47	19.70	20.40	0.70	27.76
Mary/Ahal	16.72	4.30	21.02	0.46	20.68	21.36	0.68	25.72
Nukus	11.48	4.41	15.89	0.55	15.49	16.30	0.81	38.44
Surxhondaryo	15.68	4.68	20.36	0.46	20.02	20.70	0.68	29.84
Urgench	13.39	4.36	17.75	0.53	17.35	18.14	0.79	32.55
Vaksh	17.23	4.93	22.16	0.52	21.78	22.55	0.77	28.63

Table 2.5 shows annual temperature anomalies (°C) for 2070–2099 under the A2 scenario for each of the eleven locations of interest in the Amu Darya Basin, relative to the CRU high-resolution observed data for the historical period of 1961–1990. Table 2.6 then applies the ensemble mean anomalies from the seven GCMs to the CRU 1961–1990 baseline means to provide downscaled absolute mean temperature estimates for 2070–2099 for each location.

The monthly mean modelled increases in temperature for the period 2007–2099 and the variability between the 7 models are shown in Figures 2.8 and 2.9 for Bukhara. The changes in temperature of the other locations show a similar pattern.

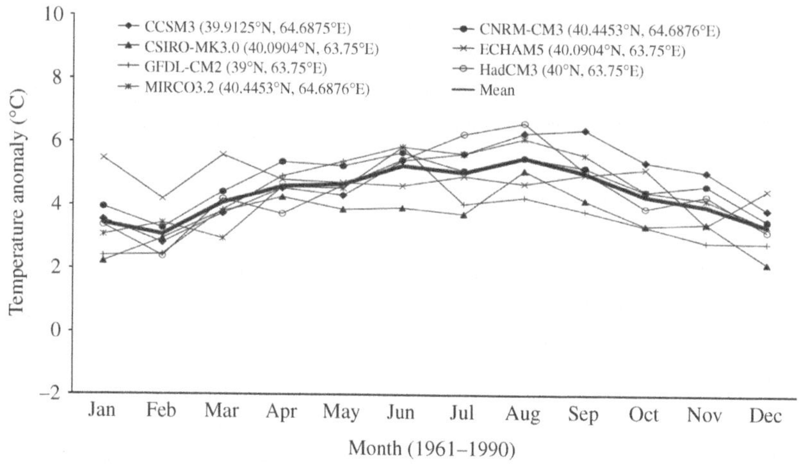

Figure 2.8. A2 30-year 2 m mean surface temperature anomalies (Jan–Dec, 2070–2099) relative to the 1961–1990 baseline normal for Bukhara.

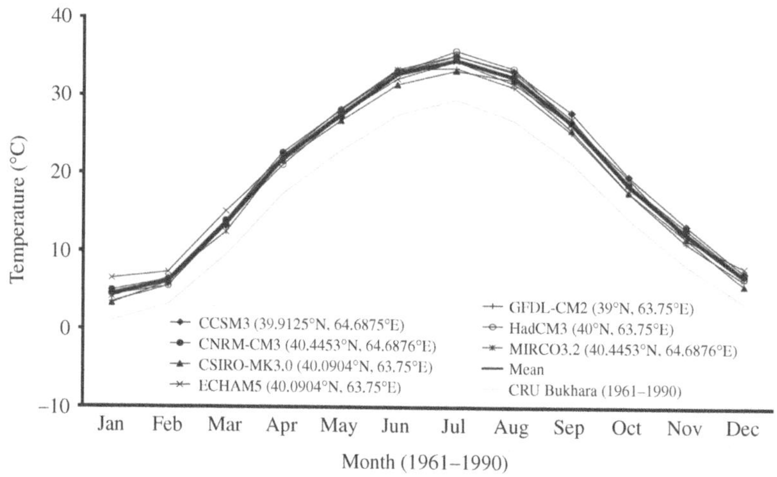

Figure 2.9. A2 30-year 2 m mean surface temperature climatologies (Jan–Dec, 2070–2099) downscaled using the 1961–1990 baseline normal for Bukhara.

Changes are also evident in the seasonality and monthly variability with summer temperature increases in the growing season being appreciably greater in summer with a mean increase in temperature of the seven climate models for May, July, August, and September of 5°C compared with an annual average mean of 4.35°C across the eleven locations.

The authors note that the climate change projections from the GCMs are however at their weakest in the high mountain regions due to the relatively coarse grid resolution of the GCMs not detecting the true elevations. Given the limitations of the GCMs at high altitudes, the projected increases in mean temperatures may be expected to be lower than those demonstrated in this study.

Bukhara and the surrounding irrigation massifs lie in the centre of the Amu Darya River Basin, and have a climate that is fairly representative of much of the irrigated parts of the basin. The downscaled temperature anomalies for Bukhara are consistent with the mean projections of the GCMs under the A2 scenario, with a projected increase of 4.85°C in mean annual temperature likely. Although there remains a high level of uncertainty with the projections up to 2099, an increase in mean summer temperature in the growing season in Bukhara and the other main irrigated areas of the Amu Darya Basin of 5°C can be expected. Under this scenario, Spring will also start earlier in the year (up to four weeks earlier) by 2099, with the onset of Autumn starting two weeks later in the year.

Precipitation: As for temperature above, Table 2.7 shows annual precipitation anomalies (mm/year) for 2070–2099 under the A2 scenario for each of the eleven locations of interest in the Amu Darya Basin, relative to the CRU high-resolution observed data for the historical period of 1961–1990. Table 2.8 applies the ensemble mean anomalies from the seven GCMs to the CRU 1961–1990 baseline norms to provide downscaled total annual precipitation estimates for 2070–2099 for each location. The monthly variations in the modelled precipitation anomalies and precipitation for Bukhara are given in Figures 2.10 and 2.11. A similar pattern was observed for the other sites.

Table 2.7. A2 scenario annual precipitation anomalies (2070–2099) relative to the observed CRU 1961–1990 baseline annual mean normals for the eleven locations across the Amu Darya Basin.

Locations across the Amu Darya Basin	AR4 GCMs (A2 Scenario, 2070–2099)								
	CCS M3 mm	CNRM-CM3 mm	CSIRO-MK3.0 mm	ECHA M5 mm	GFDL-CM2 mm	Had CM3 mm	MIROC 3.2 mm	Mean PR Anomaly mm	Range mm
Bukhara	−10.36	16.90	−1.95	4.88	−89.67	15.53	−19.86	−12.08	106.57
Dashauz	−14.47	13.11	−1.50	1.82	−40.04	−1.77	−12.68	−7.93	53.15
Down Kafirnigan	−14.70	20.26	8.78	−26.91	−182.10	−7.26	−22.77	−32.10	202.37
Dushanbe	−18.03	26.87	24.80	−38.32	−163.51	−7.26	2.54	−24.70	190.39
Karakalpakstan	−13.28	−1.74	−1.50	1.82	−25.43	−1.77	−2.39	−6.33	27.26
Lebab	−10.02	15.83	−1.95	4.88	−89.67	15.53	−11.99	−11.05	105.50
Mary/Ahal	−7.84	0.91	−26.64	2.78	−99.15	0.11	−14.84	−20.67	101.93
Nukus	−13.28	−1.74	−1.50	1.82	−25.43	−1.77	−2.39	−6.33	27.26
Surxhondaryo	−14.70	20.26	−4.02	−7.60	−182.10	−7.26	−22.77	−31.17	202.37
Urgench	−12.13	13.11	−1.50	1.82	−42.47	−1.77	−12.68	−7.95	55.58
Vaksh	−6.47	26.87	24.80	−38.32	−182.10	−7.26	2.54	−25.70	208.97

Table 2.8. A2 scenario ensemble of annual precipitation climatologies (2070–2099) downscaled using the observed CRU 1961–1990 baseline annual mean normals for the eleven locations across the Amu Darya Basin. (Standard deviations, confidence intervals and percentage change estimates calculated using monthly means.)

Locations across the Amu Darya Basin	AR4 GCM Ensemble (A2 Scenario, 2070–2099)							
	CRU (1961–1990) mm	Mean PR Anomaly mm	Mean TMP mm	Standard Deviation mm	95% Conf. Interval (Lower) mm	95% Conf. Interval (Upper) mm	95% Conf. Level mm	Change %
Bukhara	157.06	−12.08	144.98	36.69	117.80	172.16	54.36	−7.69
Dashauz	109.58	−7.93	101.64	16.90	89.12	114.16	25.04	−7.24
Down Kafirnigan	266.63	−32.10	234.53	68.25	183.97	285.09	101.12	−12.04
Dushanbe	493.09	−24.70	468.39	65.39	419.95	516.83	96.88	−5.01
Karakalpakstan	109.58	−6.33	103.25	9.66	96.09	110.41	14.32	−5.78
Lebab	138.80	−11.05	127.74	36.40	100.78	154.70	53.93	−7.96
Mary/Ahal	171.67	−20.67	151.00	36.15	124.22	177.78	53.56	−12.04
Nukus	116.88	−6.33	110.55	9.66	103.39	117.71	14.32	−5.41
Surxhondaryo	394.47	−31.17	363.30	67.86	313.03	413.57	100.55	−7.90
Urgench	109.58	−7.95	101.63	17.56	88.62	114.64	26.02	−7.25
Vaksh	328.73	−25.70	303.02	72.39	249.39	356.65	107.26	−7.82

The seven GCMs selected for use in this study demonstrate a projected drop in total annual precipitation in the Amu Darya Basin by 2099, indicating a 5% to 12% decrease compared to the 1961–1990 means. Unlike for temperature however, the projections were less uniform across the eleven locations, thought to be largely due to the inherent relative inability of the GCMs to model precipitation accurately across varying elevations.

Precipitation variability was not projected to increase significantly, and no significant change in standard deviation was noted. Winter precipitation was however shown to increase, with summer precipitation shown to decrease. Thus, according to the A2 scenario, in general, the warmer summer periods are expected to become drier.

Away from the mountains, due to the precipitation falling mainly on the irrigated massifs of the plains in the Amu Darya Basin, only a small percentage can be considered to be effective in contributing to crop water requirements. Hence, the projected 5% to 12% decrease in annual precipitation may not equate to a significant change in the demand for irrigation. However, the source of the Amu Darya Basin's water is the high mountains of the Pamirs in Tajikistan and the north-western slopes of the Hindu-Kush in Afghanistan, with >800 mm annual precipitation falling predominantly as snow at the higher altitudes. This compares with the arid steppe where 300mm of annual precipitation falls near the foothills, and notably less on the open steppe.

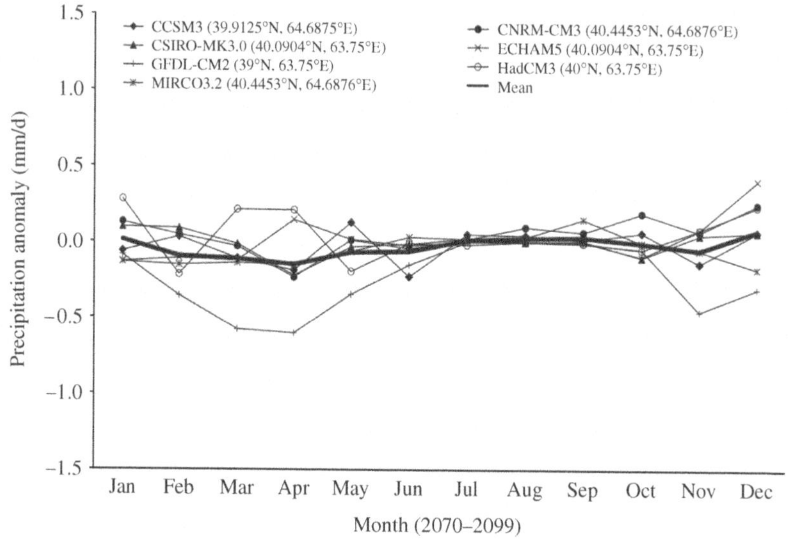

Figure 2.10. A2 30-year precipitation anomalies (Jan–Dec, 2070–2099) relative to the 1961–1990 baseline normal for Bukhara.

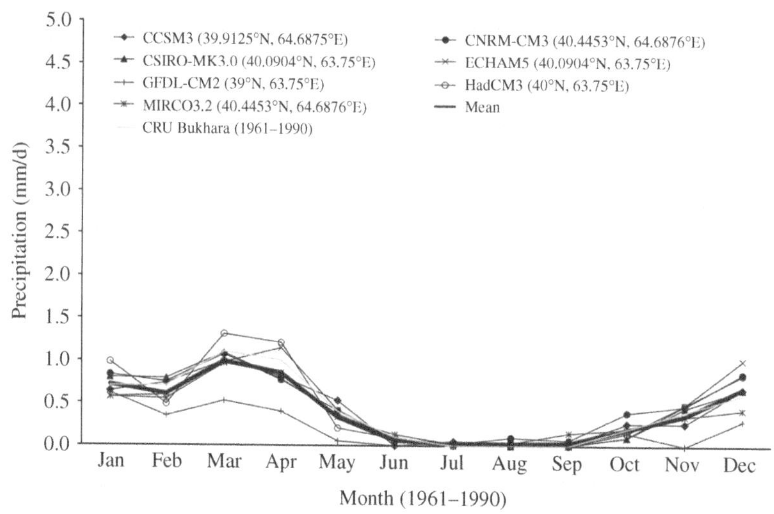

Figure 2.11. A2 30-year precipitation climatologies (Jan–Dec, 2070–2099) downscaled using the 1961–1990 baseline normal for Bukhara.

As for temperature, the GCMs are at their weakest in the high mountain regions when modelling precipitation, and hence the model projections for the region must be considered with great care. This is particularly true where localised conditions and topography play an important role in the distribution of precipitation – climate change may therefore affect the spatial distribution of precipitation with relatively different amounts falling at different altitudes and in different valleys. The snow and glacial melt of the mountains of the Pamirs and Hindu-Kush contribute to some 85% to the regions water supply, but due to lack of good hydrological data within the mountain regions, the hydrology is still poorly understood.

Although total annual precipitation is projected to decrease by 2099, the amount of runoff and its temporal distribution could however change quite appreciably. By 2099, the expected increase in temperatures may result in higher evaporative and sublimation losses. Similarly, as a result of potentially shorter winters, runoff due to snow-melt is expected to start earlier in the season and more of the precipitation in the foothills is expected to fall as rain instead of snow.

Downscaled Future Climate Projections for the Amu Darya Basin Under the A2 Scenario' for a series of monthly (January–December) data in tables and graphs for the future temperature and precipitation anomalies and climatologies for the eleven selected locations across the Amu Darya Basin for 2070 to 2099, anomalies and climatologies for 2010–2039 and 2040–2069 can also be viewed on the website address: http://www.inco-jayhun.net.

Discussion

The reanalysis of the 1961–1990 monthly means, using the CRU high-resolution observed data and the 20C3M GCM modelled data, found that the individual GCMs were relatively poor at modelling absolute climatologies for both temperature and precipitation. The decision to use the CRU observed high-resolution data as the 1961–1990 baseline normal dataset with the future projection anomalies allowing for the high variance between the GCMs to be excluded. This also enabled local downscaling to be achieved at a higher 0.5° latitude by 0.5° longitude resolution provided by the CRU dataset. The use of anomalies over absolute climatologies also allowed for the further analysis and recalculation of the future climate projections using either local observed data or a (future) improved regional observed dataset.

The range and variance shown in the future projected anomalies for temperature and precipitation from the seven GCMs was as expected. The range of values was due to the sources of error inherent in the independent development of the GCMs that has arisen from the differences and varying assumptions employed in the climate modelling of atmospheric physics, changing initial model starting conditions, computational shortfalls, and the inability to accurately predict anthropogenic economic development and its impact on the world through the various emission scenarios. It was therefore not surprising that different model outcomes at both the global and regional scale were found for the same A2 scenario.

No GCM was found to consistently model either temperature or precipitation at any location across the Amu Darya Basin better than another. In most cases, for both variables, the ensemble means of the seven GCMs were found to be the best estimate of future climate projections rather than relying on one particular GCM. When looked at as an ensemble, the anomaly projections from the GCMs provide an indication of a likely future climate and the typical range of remaining uncertainty.

It is important to note however that despite the typical range of the anomalies at the eleven locations across the Amu Darya Basin from the seven GCMs, all showed a significant rise in temperature and an appreciable decrease in precipitation at all locations up to 2099 relative to the 1961–1990 baseline period. However, the range of projections is further confused by the selection of the emission scenario. In this study, only the A2 scenario was used, which does not fully demonstrate the true variability of the projections. Although this does not give a measure of stochastic certainty, it does allow water resource planners to develop robust water resource plans for the future. In managing the risks associated with climate change on water resources it cannot be considered in isolation, but requires integration into an overall assessment of total risk.

Data sources

Data was supplied from the following sources (Table 2.9); these data sources should be reference if using the data or findings from this publication.

Table 2.9. Data sources used in this study.

Dataset	Institution/Organisation	Format
CRU 30-Year Means (0.5° grid)	Climate Research Unit, University of East Anglia, UK (hosted by IPCC-DDC)	ASCII (.txt)
CRU TS 2.1 Monthly Means (0.5° grid)	Dr. Tim Mitchell, Tyndall Centre for Climate Change Research	ASCII (.txt)
IPCC AR4 Global 30-Year Means (Summary Data)	Earth System Grid (ESG) Portal at PCMDI (WCRP CMIP3 Multi-Model Dataset) (hosted by IPCC-DDC)	netCDF (.nc)
IPCC AR4 Global Monthly Means	Modelle & Daten (CERA Database)	GRIB (.grb)

Source: Data for each site is available from the website address: http://www.inco-jayhun.net

2.3 CLIMATE CHANGE IN THE MOUNTAINOUS REGIONS OF THE UPPER AMU DARYA BASIN

2.3.1 Climate in the runoff forming zone

The south-west area of Zerafshan ridge, the south slope of Alay ridge and the Pamir area are runoff forming zones in the frame of given research. Runoff forming zones can be conditionally broken down onto mountain (lower that 3 km above sea level) regions and alpine ones. Five meteorological stations (black points) were identified as basic ones that characterize the different runoff forming areas in the interested river basins (Figure 2.12).

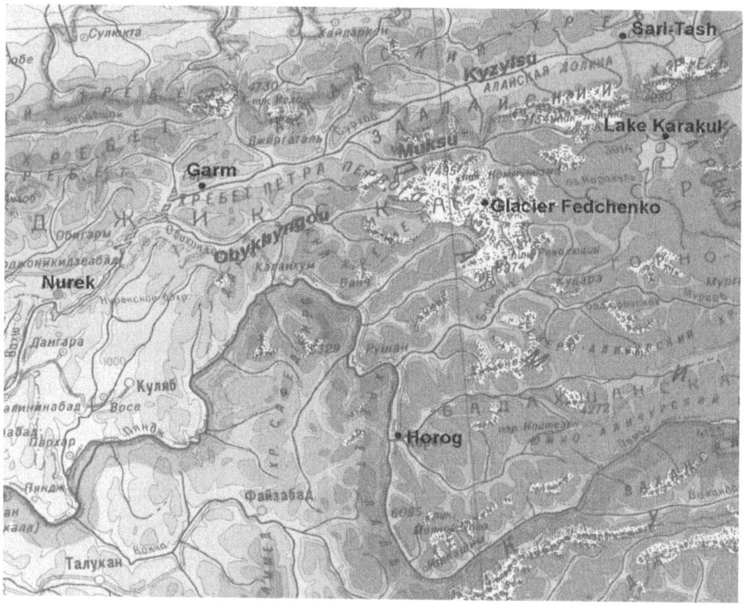

Figure 2.12. Basic meteorological stations (Garm, Horog, Sari-Tash, Lake Karakul and Glacier Fedchenko).

Quite severe winters with steady snow cover are very common for the mountain area. Annual precipitation is ranged 700–1500 mm (734 mm reported by Garm station) and is gone up with elevation (Figure 2.13). Air temperature is much low for both winter and summer time in the alpine part of the region. Annual and daily fluctuations are smoothed. Rate of precipitation is lower than in the mountain part of region (373 mm reported by Sari-Tash station).

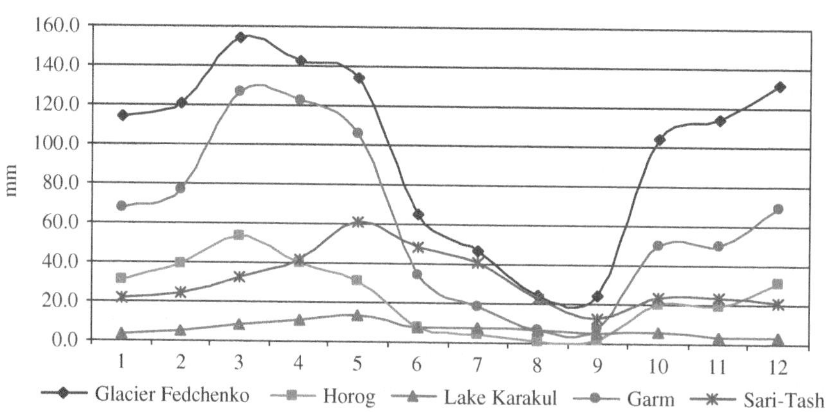

Figure 2.13. Norms (1961–1990) of the monthly precipitation for runoff forming zone by data of basic stations.

Pamir area is characterized with very irregular relief. Mountain (Horog station), alpine and glaciation areas could be distinguished there. Winter time is characterized with low monthly air temperature (Figure 2.14). Snow cover is steady for long time. Summer is quite hot. Region with low spatial air temperature fluctuations and with typical precipitation annual distribution can be divided onto dry part with annual precipitation not higher than 300 mm and wet part where annual 900 mm can be brought with precipitation being subjected to differences in aspect of slopes.

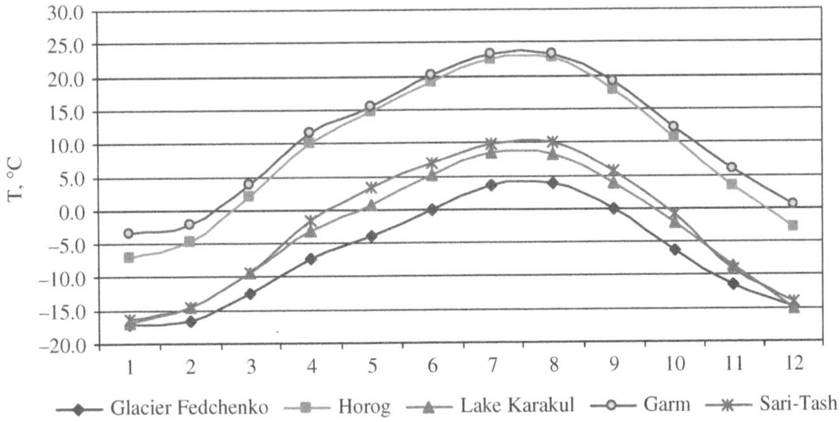

Figure 2.14. Norms (1961–1990) of the monthly air temperature for runoff forming zone by data of basic stations.

Alpine Pamir region differs from mountain one by more severe winter. Monthly air temperature can drop down −16 to 18°C in January there. Steady snow cover is kept for 4–5 months spite of low precipitation rate. Summer is cool with monthly air temperature not higher +10°C in July.

Glaciation area is characterized with low monthly air temperatures, low daily air temperature fluctuations and very long winter season with steady snow cover. Rate of precipitation is high, 1176 mm per year for Glacier Fedchenko station as instance. Maximum of precipitation comes in the spring time (Figure 2.13). Summer is cool with air temperature norm in July not higher than 4°C. There is practically no frost-free period there.

The Eastern Pamir is very dry area. Annual precipitation is lower than 80 mm (Karakul station records). Air temperature regime of the Eastern Pamir is quite similar to one of alpine region of the Alay ridge that is proved by the historical monthly air temperature time series measured on the Sari-Tash and Lake Karakul stations (Figure 2.14)

Thus, selected basic stations are representative from view point of climate characteristics of the territory in interest, they are included to the set of the basic stations that serves for runoff evaluation and can be used for analysis of climate change.

2.3.2 Current climate change in mountains part of Amu Darya Basin

Unfortunately quality of observations became worse after 1990 years because many stations stopped their operations and caused therefore a lot of gaps in the data series.

Analyses have shown that climate of mountains part of Amu Darya Basin is affected by global warming. Temperature increase is clearly observed virtually in all areas and altitude zones (from +0.3°C to +1.3°C, Figure 2.15) for the last 60–70 years.

Alongside with air temperature changes, the precipitation character changes have also been observed. Annual variations of precipitation are very high. The changes of a standardized index of precipitation (the annual sums of precipitation relative to 1961–1990 norms) are shown in Figure 2.16. Weak tendencies of increase in precipitation are observed on the majority of stations. Trends increments depending on territory make up 0–15%.

2.3.3 Climate projection for mountains part of Amu Darya Basin

Climate scenarios for mountain part of Amu Darya Basin are based on General Circulation Models (GCMs) and the MAGICC/SCENGEN 4.1 software. MAGICC (Model for the assessment of Greenhouse-gas Induced Climate Change) consolidates information on regional climate changes structure that has been acquiring by different GCMs with outputs of the "simple" climate model. The climate model in MAGICC is an upwelling-diffusion energy-balance model that produces global – and hemispheric-mean output. Global-mean temperatures from MAGICC are used to drive SCENGEN (Scenario Generator). SCENGEN uses the scaling method produce spatial patterns of change from an extensive data base of GCMs. The pattern scaling method is based on the separation of the global-mean and spatial-pattern components of future climate change, and the further separation of the latter into greenhouse-gas and aerosol components. Spatial patterns in the data base are 'normalized' and expressed as changes per 1°C change in global-mean temperature. These normalized greenhouse-gas and aerosol components are appropriately weighted, added, and scaled up to the global-mean temperature defined by MAGICC for a given year and emissions scenario. SCENGEN allows selecting and averaging outputs of different GCMs for the patterns of greenhouse-gas-induced climate for particular region (Wigley, 2003).

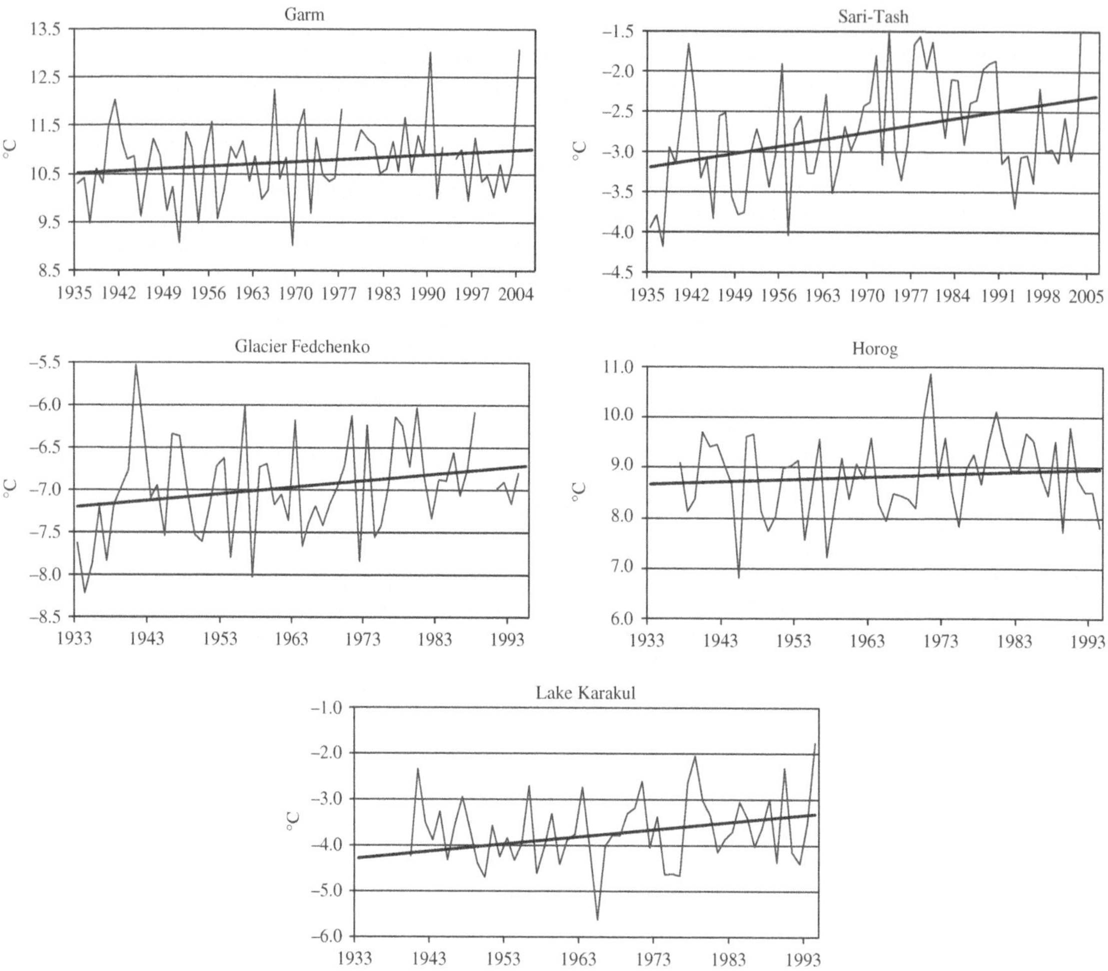

Figure 2.15. Annual temperature change in the mountains part of Amu Darya Basin (°C).

With the purpose to build regional climate scenarios for mountain part of Amu Darya Basin, middle climate sensitivity to growth of greenhouse gases concentration in the atmosphere was selected. All range of emission scenarios was considered. The results by six models were averaged (HadCM3, UK, Hadley Centre; ECHAM4, Germany, Max Planck Institute; CSIRO-TR, Australia, Scientific Industrial Research Organization; GFDL-TR, USA, Geophysical Fluid Dynamics Laboratory; CGCM1 – TR, Canadian Climate Centre; CGCM1 – TR, Canadian Climate Centre; CCSR-NIES, Japan.). Published results on analysis the modelling for some regions in the world availability, recommendations on models selection under designing the scenarios of climate changes impact given in (IPCC, 2007; Parry, 2002; Spectorman, 2002; Spectorman and Petrova, 2007) were used as input for decision making on models to be selected. Moreover the experiments carried out during first year of project have shown that averaging the results of several models would partly moderate the uncertainty of the climate scenarios designed for mountain territories of Central Asia.

Differences between models are indicator of the credibility of climate projections, and inter-model differences can provide a quantitative measure of the uncertainty in these projections. Below (Table 2.10) there are the comparative data with signal-to-noise ratio (SNRs) as criterion that is calculated in the SCENGEN as ratio the expected change at the end of century (under yearly increase of greenhouse gases concentration for 1%) to inter models differences (Wigley, 2003).

Analysis of results given in Table 2.10 shows:

- uncertainty the precipitation scenarios is higher then air temperature scenarios;
- procedure of averaging gives higher SNR in most study cases;
- some an increase in precipitation rate is expected at the end of century for DJF season;
- increase and decrease in precipitation rate is equiprobable for spring-fall season;
- decrease in precipitation rate is expected over the all mountain territory of Central Asia for summer time.

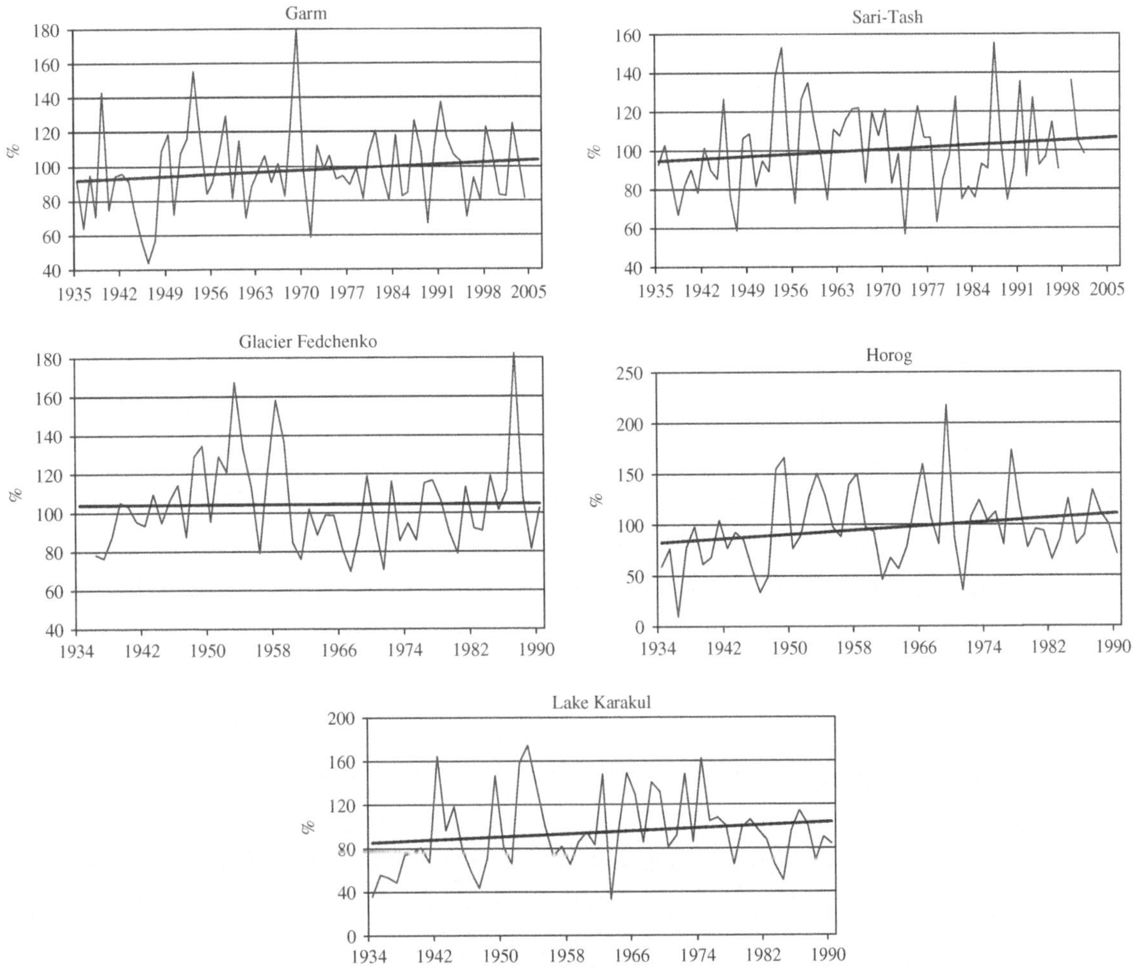

Figure 2.16. Change of standardized precipitation index (%) in the mountains part of Amu Darya Basin.

Table 2.10. Inter-model signal-to-noise ratio representatives for mountain area of Central Asia.

Points situation			**6 selected models**				**All 17 models, available in SCENGEN**			
N	**Lat**	**Lon**	**DJF**	**MAM**	**JJA**	**SON**	**DJF**	**MAM**	**JJA**	**SON**
Temperature										
1	42.5	67.5	3.67	4.12	4.24	6.75	3.63	3.62	4.61	7.09
2	42.5	72.5	4.1	5.01	4.22	6.87	4.43	4.13	4.51	7.01
3	37.5	67.5	3.11	4.27	4.75	5.95	3.04	4.08	5.19	6.14
4	37.5	72.5	5.28	5.92	4.68	7.63	4.46	4.48	5.33	7.06
Precipitations										
1	**42.5**	**67.5**	**0.46**	**−0.1**	**−1.22**	**0.07**	**0.33**	**−0.4**	**−0.88**	**−0.03**
2	**42.5**	**72.5**	**0.85**	**0.34**	**−0.67**	**0.16**	**0.7**	**0.01**	**−0.68**	**0.26**
3	**37.5**	**67.5**	**0.46**	**−0.29**	**−2.68**	**−1.01**	**0.33**	**−0.56**	**−0.83**	**0.23**
4	**37.5**	**72.5**	**0.89**	**−0.03**	**−0.96**	**−0.9**	**0.63**	**−0.16**	**−0.97**	**0.17**

Downscaling of the climate projection

Climate scenarios have been designed in accordance with emission scenarios (A1B, A1FI, B1, A2, B2) with taking into consideration of sulphate aerosols influence for three time intervals in the future (2016–2045, 2036–2065, 2066–2095) that would characterize of climate changes by 2030, 2050 and 2080. Statistical method being based on multivariate linear regression was applied for getting the detailed information over the territory (Spectorman, 2002).

This approach to developing the statistical downscaling method is the 'ideal forecast' concept, when statistical links are found in the diagnostic data but applied to the GCM output data. In this case the quality of interpretation (downscaling) improves with the improvement of the models. To these end archives of each climatic parameter should be created in the grid nodes based on the monitoring data that should be viewed as predictors, while the actual monitoring data obtained at stations should be viewed as predictants.

A specialised archive of predictors has been created in the grid nodes based on the monthly resolution data supplied by all the stations existing in the territory under review.

Detailed elaboration of scenarios was implemented by data came from stations located in the runoff forming zone of Amudarya (Table 2.11). Twenty-two stations were available to proceed downscaling on air temperature data and 21 stations on precipitation data. Mountain stations mostly located on territory of Tajikistan.

Table 2.11. List of meteorological stations located in the runoff forming zone of the Amu Darya.

N	Station	Latitude	Longitude	Elevation, m.
1	Dehavz	39° 27′	70° 12′	2564
2	Dushanbe-agro	38° 33′	68° 48′	791
3	Iskanderkul	39° 06′	68° 23′	2196
4	Kulyab	37° 55′	69° 47′	612
5	Kurgan-Tube	37° 49′	68° 47′	427
6	Hudjant	40° 13′	69° 44′	410
7	Ura-Tube	39° 54′	68° 59′	1005
8	Glacier Fedchenko	38° 50′	72° 13′	4169
9	Khorog	37° 30′	71° 30′	2077
10	Lake Karakul	39° 01′	73° 33′	3932
11	Anzobskiy Pass	39° 05′	68° 52′	3375
12	Garm	39° 00′	70° 18′	1317
13	Dangara	39° 06′	69° 19′	660
14	Isfara	40° 08′	70° 36′	847
15	Shakhristanskiy Pass	39° 34′	68° 35′	3143
16	Sangiston	39° 23′	68° 35′	1502
17	Tavil-Dara	38° 42′	70° 29′	1617
18	Madrushkent	39° 26′	69° 40′	2253
19	Lyakhsh	39° 17′	71° 32′	2002
20	Saritash	39° 43′	73° 15′	3155
21	Maykhura	39° 02′	68° 47′	1920
22	Fayzabad	38° 32′	69° 19′	1216
23	Ishkashim	36° 62′	71° 73′	2524

The "Ideal forecast" archive was set up in the nods 5×5° that cover foothill and mountain territories of Central Asia (points 1,2,3,4 see Table 2.10). Data from Uzbekistan and Kyrgistan were additionally involved to the archive setting up. Such archive was completed for 1951–1991 time period, as period with most reliable and available data set.

Statistical characteristics of "ideal forecast" for air temperature in the 5×5° nods, calculated for mentioned above time period, are given in the Table 2.12. The characteristics of inter-points spatial-correlation relationships are given in the Table 2.13.

Table 2.12. Statistical characteristics of monthly air temperatures in the 5×5° nods for central months in seasons.

Point number	January Mean	SD	April Mean	SD	July mean	SD	October mean	SD
1	−3.4	2	11.3	1.6	22.8	1.2	10.8	1.5
2	−11.5	2.5	10.4	1.6	18.8	1.4	4.9	1.7
3	1.1	2.2	13.7	1.3	23.9	0.8	14	1.5
4	−9	1.4	11.6	1.5	17.8	1.2	6	1.2

Analysis of the spatial correlation coefficients for monthly air temperature showed there is no significant correlation between air temperature in Central Pamir (point 4) and other points for winter time whereas correlation goes up for another seasons.

Table 2.13. Correlation coefficients between points of "ideal forecast" archive for air temperature.

January					July				
N	1	2	3	4	N	1	2	3	4
1	1.00				1	1			
2	0.92	1.00			2	0.86	1		
3	0.88	0.80	1.00		3	0.76	0.65	1	
4	0.21	0.34	0.28	1.00	4	0.64	0.66	0.60	1
April					October				
N	1	2	3	4	N	1	2	3	4
1	1.00				1	1			
2	0.79	1.00			2	0.88	1		
3	0.71	0.57	1.00		3	0.73	0.71	1	
4	0.71	0.60	0.71	1.00	4	0.67	0.67	0.65	1

This fact points out that there is correlation between predictors and way to calculate the multivariate equations (3 predictors) for relationship between air temperature in the nods and data come from stations only for winter time. Univariate equations had been calculated for another seasons. Spatial relationships for monthly precipitation sums in $5 \times 5°$ nods appeared as insignificant and that fact gave grounds to use 3 predictors to design calculated equations.

Relationships between monthly air temperatures on stations (predictants) and predictors (ideal forecasts – average anomalies) are very close. For example, values of summary factors of correlation in the majority of months of year 0.84–0.97. Relationships between monthly precipitations are lower, but statistically significant, that allows to use them for statistical interpretation of GCMs results.

Thus calculated equations for air temperature and precipitation by scenarios were designed for each station and month listed in the Table 2.11. This approach succeeded in downscaling of monthly scenarios from nods of calculated network to mountain station listed in the Table 2.11.

Decision on emission scenarios

The increase of the air temperature for three time intervals (by 2030, 2050 and 2080) was calculated with help of MAGICC model for both whole globe and Central Asia territory. The results given in the Table 2.14 point out the only particular results can be taken into consideration due to insignificant bias between some of expected warming rates and even there are practically nothing differences between some of scenarios. That's why more reasonable and sufficient way is to consider three scenarios that would characterize very strong, strong and moderate options of climate warming to get assessment of climate change impact to the runoff forming zone.

Table 2.14. Expected yearly air temperatures by different scenarios for greenhouse gases emission.

Emission scenarios	**Globe warming rates**			**Warming rates for mountainous part of Central Asia (point 4)**		
	2030	**2050**	**2080**	**2030**	**2050**	**2080**
A1FI	0.75	1.7	3.4	1.5	3.1	5.9
A1B	0.8	1.5	2.4	1.6	2.6	4.5
A2	0.72	1.25	2.4	1.4	2.2	4.3
B2	0.8	1.3	2.0	1.7	2.5	3.8
B1	0.9	1.4	1.8	1.8	2.6	3.6

However, scenarios data calculated for particular stations are to be analyzed with data derived via statistical interpretation and current climatic conditions are to be taken into consideration to identify the basic stations to make decision on scenarios are to be selected.

Analysis of the seasonal air temperature and seasonal precipitation fluctuations expected by the different scenarios

The expected seasonal fluctuations of air temperature calculated for three time periods (by 2030, 2050, 2080) with data from basic stations given in Figures 2.17 to 2.20. The analysis indicates that to consider only three emission scenarios for climate changes impact are a sufficient amount for the runoff assessment. The expected changes within group B

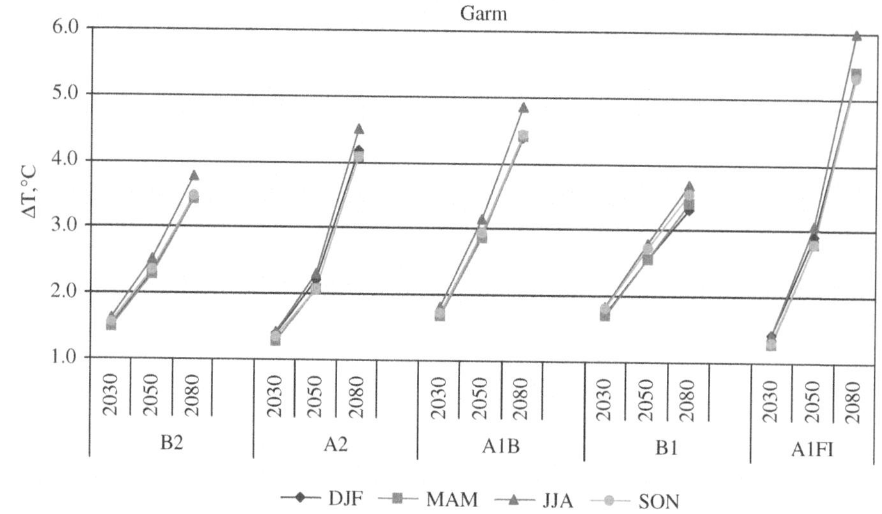

Figure 2.17. Expected seasonal air temperature changes per different scenarios and time periods (Garm station).

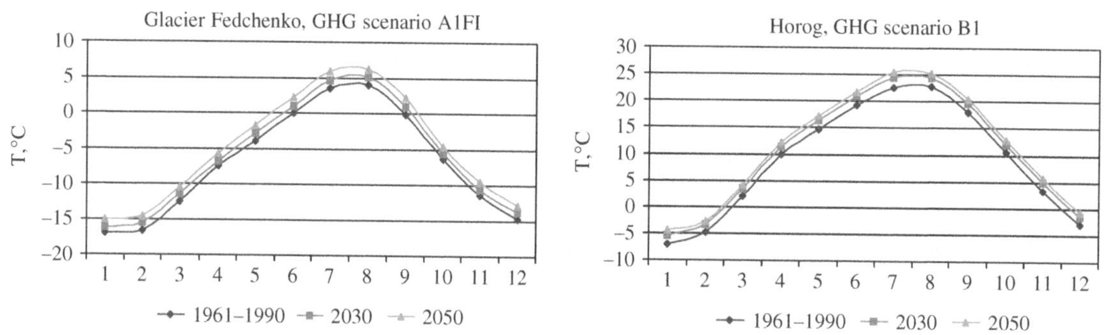

Figure 2.18. Annual air temperature fluctuations for basic time period and expected ones per climate scenarios for Glacier Fedchenko and Horog.

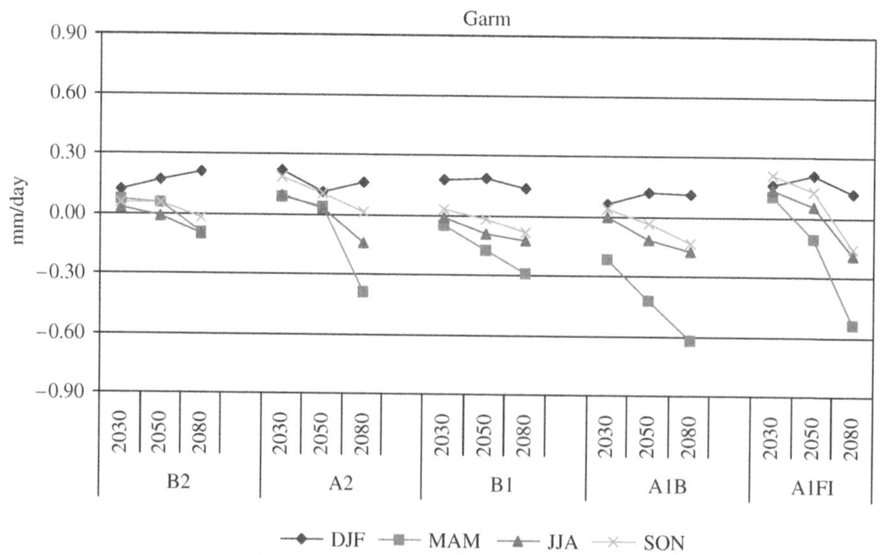

Figure 2.19. Expected seasonal precipitation changes per different climate scenarios and time periods for Garm station.

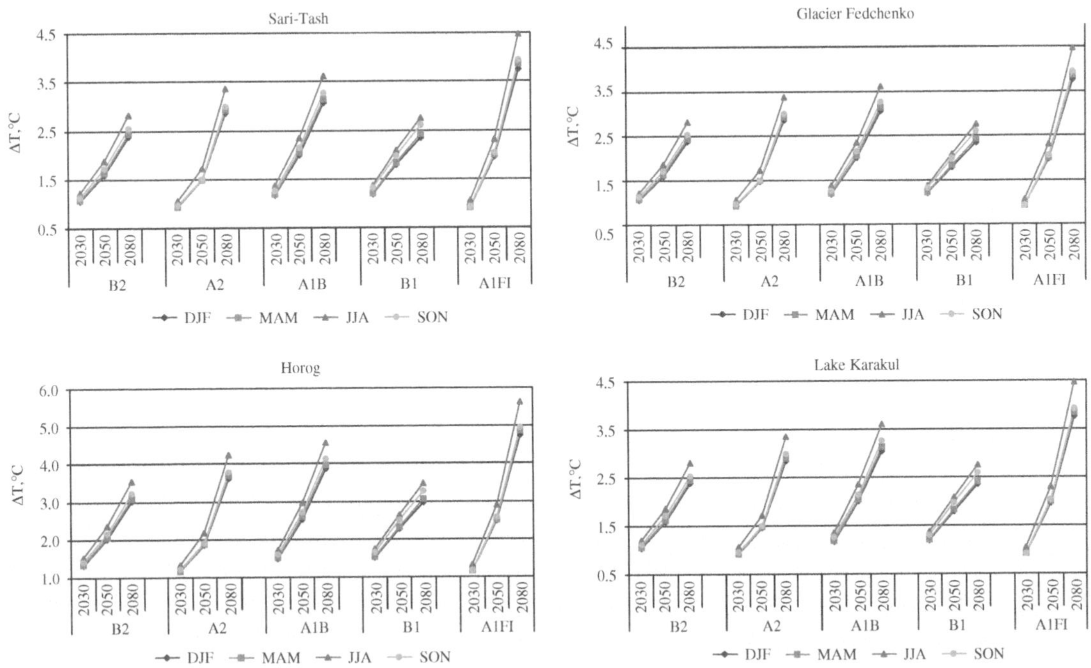

Figure 2.20. Expected seasonal air temperature changes per different climate scenarios and time periods for basic stations.

scenarios are close each other and there is an extreme scenario A1FI in the group A but it will be expected for far perspective. The highest rates of warming are expected by 2030 per A1B scenarios and warming with minimal rate is expected for A1FI. The minimal air temperature changes are expected for A2 scenarios and maximal ones are expected for A1B scenarios (Figures 2.17–2.20).

Warming tendency keeps more intensive way for foothill area than for alpine one. Summer air temperature increase up to 6° is expected for extreme scenarios A1FI by 2080 at Garm station (1317 m.a.s.l.) whereas increase up to 4.4° is expected for Glacier Fedchenko station (4169 m.a.s.l.); all these follow the actual tendencies now: warming rate for alpine regions is lower than rate for mountain and foothill ones.

The even increase of air temperature is recognized for all scenarios under consideration of the expected warming month to month. Annual air temperature fluctuations for basic time period and expected ones per climate scenarios A1F1, B1 are shown in Figure 2.18 as example.

Expected precipitation changes are less agreed (Figures 2.19–2.21) under consideration of their absolute changes (mm/day). However, the following features can be noted when expected precipitation changes are expressed as rate of the base norm (Table 2.15):

- the low precipitation increase (up to 10%) is expected for winter time;
- more sensitive to the warming process are spring, summer and fall seasons. Per all scenarios the precipitation decrease (up to 16% for spring and fall, 30% for summer) is expected for far perspective;
- the most climate aridity will have to be expected per A1B scenarios for short-term perspective (by 2030). Per scenarios given some decrease of annual precipitation is expected. Lowest rate of climate aridity is expected per scenarios A1F1 with annual precipitation decrease rate 5–8%;
- the climate aridity will become more severe in the investigated mountain region per all scenarios for far perspective (by 2080).

Conclusions

Considering the specific interest of water resources assessment for short-term perspective and results of precipitation and temperature scenarios the following could be concluded:

- in temperature regime the difference between A(A1B, A2) and B(B1, B2) groups of scenarios is insignificant until 2050;
- at least 3 scenarios are to be considered: A1B or A2 (moderate rate of temperature increase to 2080), A1F1 (the highest rate of temperature increase), B1 or B2 (low rate of temperature increase).

Thus, the high degree of water resources vulnerability to climate changes has been shown through the expected high rate of warming and expected decrease of precipitation especially for long term perspective. Risk of strong spring floods is much anticipated with expected precipitation increase for winter time under intensive warming tendency.

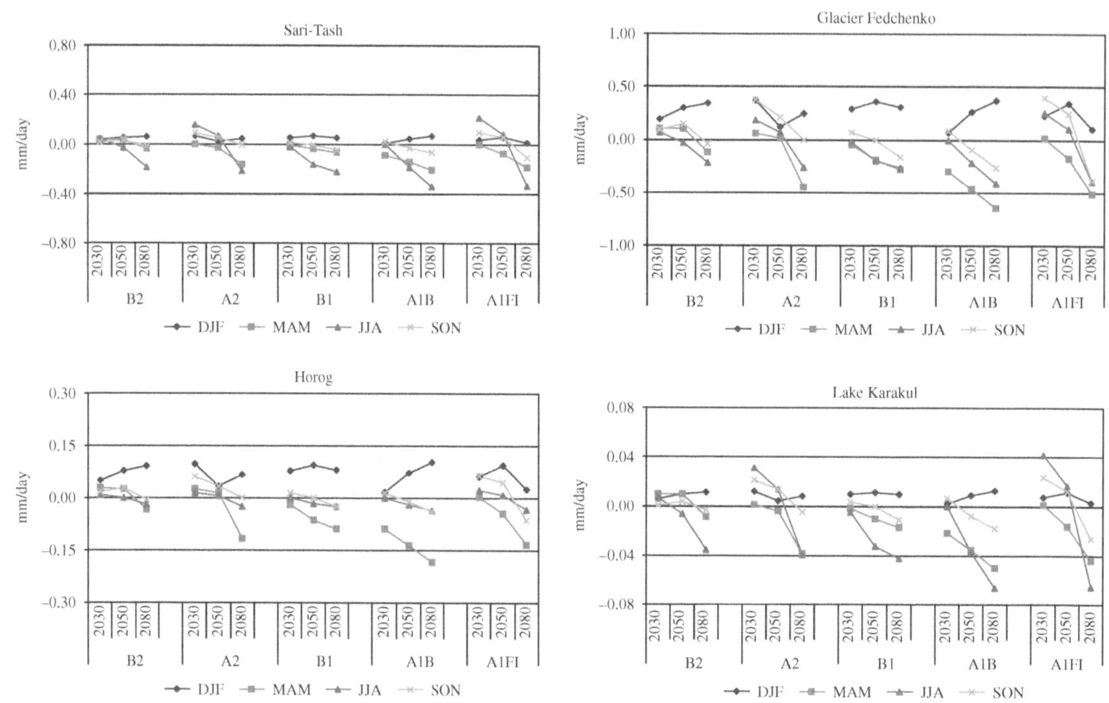

Figure 2.21. Expected seasonal precipitation changes per different climate scenarios and time periods for basic stations.

The designed climate scenarios with monthly resolution of 23 stations located in the runoff forming zone for calculated time periods (2030, 2050, 2080) per 3 emission scenarios (A1B, A1FI, B1, A2, B2) are principle results. Scenarios data base (available in Excel) consist of climate norms (1961–1990), scenarios (expected changes) as both rates of norms (air temperatures and precipitation anomalies as rate of norms%) and expected air temperatures and precipitation derived per scenarios of greenhouse gas emission. The designed climate scenarios are informational platform to get assessment on runoff changes in runoff forming zone with mathematical modelling approach.

2.3.4 Estimation of extreme values of temperature and precipitation for climate scenarios

Extremeness of air temperature regimes in the future characterises the temperature values and the precipitation sums of the specified probability (percentiles p5, 10p, 90 and 95p) for selected climatic scenarios.

The design procedure is based on the use of theoretical functions of probabilities distribution. For air temperatures the normal distribution with correction for asymmetry is used and air temperatures series in winter months are characterized by negative asymmetry. For the monthly sums of precipitation Pearson curve of type III is used. The calculation of values for the specified probability is a standard procedure if the basic statistical characteristics (average and standard deviations) are known. The constructed climatic scenarios allow assessing changes of average values for the various time periods.

Estimations of standard deviations changes have been done using MAGICC/SCENGEN software according to different emission scenarios. Estimations have shown the strong growth of variability of air temperature and weak growth of precipitation variability concerning the modern period on considered territory (Figure 2.22). Concerning the asymmetry factors the assumption of their conformity to empirically designed values has been made.

The expected increase in standard deviations specifies strengthening of extremeness for a regime of air temperatures in the future, i.e. increasing the probability of extreme hot weather conditions and preservation of high probability of extremely low air temperatures even on a background of intensive warming.

Variability of air temperature and precipitations will grow according to warming rates. Extreme scenario A1FI shows the increase in variability more than on 60% by 2080. The increase in precipitation variability in percentage expression is less significant, but, taking into account the current high variability, increase in a range up to 10% also represents serious threat.

Corresponding percentiles for all stations of a considered mountains zone are calculated using designed parameters of probabilities distribution (average values and standard deviations). On Figure 2.23 the designed values of a median (50p) and percentiles 10p and 90p of monthly temperatures and the sums of precipitation for Glacier Fedchenko station are submitted, which allow to estimate a possible range of fluctuations of temperatures of air and precipitations in the future.

Table 2.15. Expected monthly precipitation (% rate of base norm calculated for 1961–1990) for particular station located in the Amu Darya runoff forming zone.

Scenarios	Station	Winter			Spring			Summer			Fall			Year		
		2030	2050	2080	2030	2050	2080	2030	2050	2080	2030	2050	2080	2030	2050	2080
B2	Glacier Fedchenko	4.7	7.3	8.4	2.2	2.1	−2.5	4.3	−2.0	−14.5	3.4	5.4	−1.6	3.4	4.0	−0.3
	Horog	4.5	7.0	8.2	1.9	1.9	−2.5	6.0	−0.9	−13.9	4.0	6.1	−1.5	3.3	4.2	1.0
	Lake Karakul	4.4	6.9	8.1	2.6	2.6	−2.5	2.8	−2.7	−15.0	0.5	2.4	−2.0	2.6	1.8	−4.1
	Garm	4.9	7.2	8.8	1.7	1.3	−2.4	4.5	−1.7	−15.9	4.8	4.7	−2.0	3.3	3.3	−0.2
	Sari-Tash	4.6	7.2	8.3	2.6	2.7	−2.5	3.5	−2.6	−14.9	2.3	4.2	−1.7	3.2	2.1	−4.2
A2	Glacier Fedchenko	9.0	3.0	5.9	1.2	0.2	−9.3	12.0	4.7	−17.6	14.0	7.7	−0.2	7.5	3.1	−3.6
	Horog	8.7	2.9	5.8	1.7	1.0	−8.5	10.9	3.7	−18.3	14.0	7.6	0.2	6.4	2.8	−2.5
	Lake Karakul	8.6	2.9	5.8	0.3	−1.1	−10.6	13.2	5.6	−16.7	13.4	8.5	−3.0	7.3	2.9	−8.4
	Garm	9.2	4.7	6.8	2.3	0.9	−10.0	14.8	3.8	−21.9	15.3	8.5	1.3	7.2	3.3	−4.4
	Sari-Tash	8.8	2.9	5.9	−0.1	−1.7	−11.0	12.3	5.1	−17.4	13.9	8.0	−1.1	7.4	2.7	−8.3
B1	Glacier Fedchenko	7.1	8.9	7.5	−1.1	−3.9	−5.8	−1.2	−13.1	−17.7	2.4	−0.3	−6.0	2.2	−0.2	−3.1
	Horog	6.9	8.5	7.2	−1.4	−4.6	−6.5	0.0	−12.7	−17.5	2.9	0.2	−5.5	2.3	0.5	−1.8
	Lake Karakul	6.8	8.2	7.1	−0.5	−2.8	−4.8	−2.2	−13.6	−17.9	2.2	−0.4	−7.1	0.6	−3.5	−6.8
	Garm	7.6	7.7	5.7	−1.4	−4.4	−7.3	−2.1	−14.5	−19.8	2.4	−1.4	−7.1	1.7	−1.3	−4.5
	Sari-Tash	7.1	8.8	7.3	−0.3	−2.5	−4.4	−1.8	−13.2	−17.7	2.1	−0.6	−6.6	0.9	−3.4	−6.6
A1B	Glacier Fedchenko	1.6	6.4	9.1	−6.2	−9.8	−13.4	0.0	−14.7	−27.3	3.2	−3.4	−9.6	−1.1	−4.0	−7.2
	Horog	1.5	6.4	9.3	−6.4	−9.9	−13.2	0.0	−13.7	−26.6	3.3	−2.4	−8.9	−1.8	−3.1	−5.0
	Lake Karakul	1.5	6.3	9.2	−5.9	−9.6	−13.7	0.0	−15.6	−27.9	4.2	−5.4	−11.5	−1.4	−8.0	−13.6
	Garm	2.6	5.1	4.8	−5.5	−10.8	−15.9	−0.5	−17.7	−27.5	2.9	−3.0	−11.0	−1.5	−5.6	−10.1
	Sari-Tash	1.4	6.5	9.4	−5.8	−9.5	−13.6	0.0	−15.0	−27.6	3.4	−4.4	−10.5	−1.3	−7.5	−13.2
A1FI	Glacier Fedchenko	5.6	8.5	2.6	0.3	−3.7	−10.7	17.0	6.6	−26.2	14.7	9.2	−14.5	6.8	3.9	−9.1
	Horog	5.4	8.4	2.2	0.2	−3.2	−9.7	16.4	6.4	−24.7	14.5	9.7	−14.3	4.8	3.3	−6.7
	Lake Karakul	5.4	8.3	1.9	0.3	−4.4	−12.1	17.4	7.0	−27.6	15.0	7.5	−17.1	8.2	2.6	−14.9
	Garm	6.8	9.0	4.9	2.6	−2.7	−13.7	20.8	8.2	−29.2	17.4	11.0	−13.5	7.5	3.6	−9.5
	Sari-Tash	5.2	8.4	2.4	0.0	−4.8	−12.3	17.2	6.6	−26.7	14.8	8.5	−15.3	8.4	3.1	−14.4

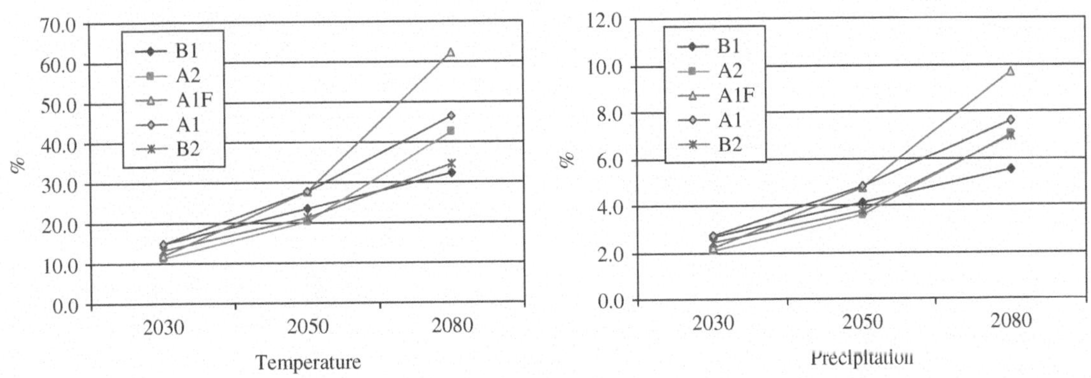

Figure 2.22. Estimation of standard deviations changes of annual air temperature in a mountains zone according to different emission scenarios (point with coordinates 37.5°N, 72.5°E).

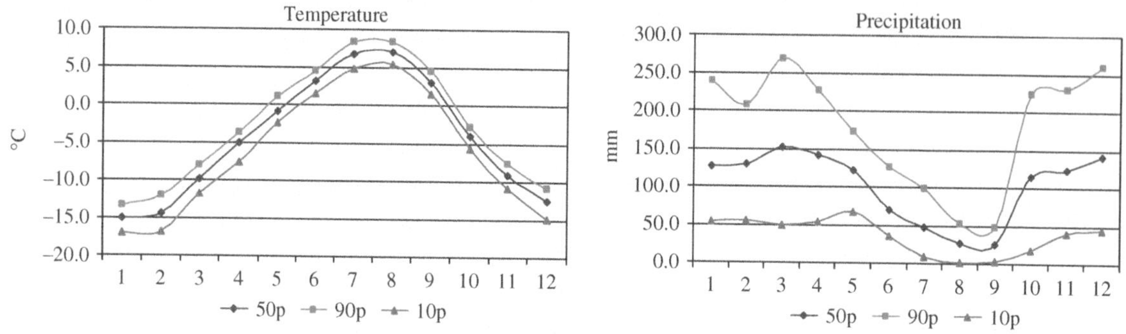

Figure 2.23. Designed values of a median (50p) and percentiles 10p and 90p of monthly temperatures and the sums of precipitation for Glacier Fedchenko station according to A2 scenario.

The constructed climatic scenarios for all stations where were available long time series of observation (on air temperature was possible to use 22 stations, on precipitation – 21) are an information basis for an estimation of a runoff changes from a zone of formation with application of mathematical modelling.

Scenarios are received for the future time periods (2030, 2050 and 2080) according to 5 green house gases scenarios (A1B, A1FI, B1, A2, B2). Scenarios include base norms (1961–1990), expected changes, as in relative expression, and actually values of temperature and precipitation, and also the designed sizes of the specified probability which are a basis for an estimation of potential variability, extremely high and extremely low runoff in Amu Darya Basin.

2.4 REFERENCES

IPCC (2007) Regional climate projections. In *Climate Change 2007: The Physical Science Basis. Contribution of Working Group I to the Fourth Assessment Report of the Intergovernmental Panel on Climate Change*. Solomon, S., Qin, D., Manning, M., Chen, Z., Marquis, M., Averyt, K.B., Tignor, M. and Miller, H.L. (eds.). Cambridge University Press, Cambridge, United Kingdom and New York, NY, USA.

New, M., Hulme, M. and Jones, P.D. (1999) Representing twentieth century space-time climate variability. Part 1: development of a 1961–90 mean monthly terrestrial climatology. *Journal of Climate*, **12**, 829–856.

Parry, M. (2002) Scenarios for climate impact and adaptation assessment. *Global Environmental Change*, **12**(2), pp. 149–153.

Patz, J.A., Hulme, M., Rosenzweig, C., Mitchell, T.D., Goldberg, R.A., Andrew K., Githeko, A.K., Lele, S., McMichael, A.J. and Le Sueur, D. (2002) Regional warming and malaria resurgence. *Brief Communications to Nature*, **420**, 627–628.

Shibuo, Y., Jarsjo, J. and Destouni, G., (2007) Hydrological responses to climate change and irrigation in the Aral Sea drainage basin. *Geophysicl Research Letters*, **34**, L21406.

Spektorman, T.U. (2002) Methods of development of Climate Change scenarios on the territory of Uzbekistan using "ideal forecast" concept. *Assessment of Climate Change on the Territory of Uzbekistan, Development of Methodological Approaches of Environmental Vulnerability Assessment*: *Bulletin*. Vol. 5, Tashkent, SANIGMI (in Russian), pp. 83–88.

Spektorman, T.U. and Petrova, E.V. (2007) Climate Scenarios on the territory of Uzbekistan. *Climate Scenarios, Climate Change Impact Assessment*: *Bulletin* Vol. 6, Tashkent: NIGMI (in Russian), pp. 14–21.

Wigley, T.M.L. (2003) MAGICC/SCENGEN 4.1: Technical Manual, National Center for Atmospheric Research, Colorado, USA, October 2003.

Chapter 3

Impact of climate change on the run-off formation zone

Mikhail Bolgov, Natalya Agaltseva, Pierre Chevallier, Bernard Pouyaud, Marina Trubetskova and Marie Mojaisky

3.1 GENESIS AND DYNAMICS OF THE RUNOFF

Water resources of the Amudarya river basin are formed in the mountains due to melting seasonal snow cover and glaciers. Climate warming worsens snow accumulation conditions in mountains and intensifies glaciation degradation.

Glaciers of the mountain part of Central Asia are the most important source and long-term reserve of clear fresh water. However ice resources are not stable. At present glaciologists notice general glacier contraction: small glaciers disappear but large ones split up.

The basic part of glaciation inside of the Aral Sea Basin is located within of watersheds Vakhsh and Panj rivers which are the main components of Amudarya. Current distribution of glaciers in the upstream of Amudarya illustrates Figure 3.1 which was prepared after processing remote sensing images for Pamir and Hindukush from satellites LANDSAT 7 and TERRA.

Data on long term change of glaciers size are necessary for modelling and forecasts of water resources in the Amudarya river basin. Total area of glacier and its morphological components (accumulation and ablation fields, bare ice and moraine cover) are main characteristics for hydrological computations and analyzing relationship between climate change and fluctuation of glaciers size. At present the reference books (Inventory 1971–1978, Schetinnikov, 1997) are sources of data on change of glaciers size related to two temporal cuts. Those are approximately end of 50-ies and 1980 in the Vakhsh River and right tributaries of Panj river basins and only 1960 for the left tributaries of Panj river (Ivan'kov, 1970).

Large gap of information for the considered area is filled now both computational methods and processing remote sensing images of Pamir and Hindukush mountain areas obtained at 2000–2001 from LANDSAT 7 and TERRA satellites. Methodical approaches used at this stage of researches are presented in publications (Agaltseva and Konovalov, 2005; Konovalov and Williams, 2005; Konovalov and Desinov, 2007) and briefly described below.

After combining data of Glacier Inventories, computational results, using remote sensing images and application of GIS techniques, the author got at the first time data on fluctuation of glaciers size in the upstream of Amudarya during 1960–2000 years. These data will be used in frameworks of Jayhun Project for substantiated modelling of glaciers runoff and influence of future climate change on water resources in the upstream of Amudarya river.

The above mentioned data on fluctuation of glaciers area in the main tributaries of Vakhsh and Panj river in 1961–2000 presented in the Table 3.1.

Figure 3.1. Glaciers in the upstream of Amudarya river basin. 1 – Amudarya river, 2 – Vakhsh river, 3 – Panj river, 4–15 are tributaries of Panj river (4 – Kunduz, 5 – Kufab, 6 – Sheva, 7 – Vanch, 8 – Yazgulem, 9 – Bartang and Murgab, 10 – Gunt, 11 – Shahdara, 12 – right and left tributaries of Panj below confluence of Pamir and Vahandarya rivers, 13 – Pamir, 14 – Vahandarya, 15 – Kudara), 16–20 are tributaries of Vakhsh river (16 – Obihingou, 17 – Surhob, 18 – Muksu, 19 – western Kyzylsu, 20 – right tributaries of Piandj below mouth of Vanch river).

Computations of monthly and seasonal glaciers runoff within Vakhsh and Panj river basins in the past and future will be based on using long term range of meteorological data. In the considered situation the set of initial information should include: monthly values of precipitation, mean, minimal and maximum air temperature, cloudiness and vapour pressure in the air. Taking into account availability of meteorological data in the past, at present and especially in future it was decided to prepare regional data base included values of listed variables in the centre of grid, having size 0.5 degree along latitude and longitude. Initial data for our regional data base were extracted from Tyndall Centre for Climate Change Research, School of Environmental Sciences, University of East Anglia, Norwich, NR4 7TJ, UK (t.mitchell@uea.ac.uk). Duration of meteorological time series in the regional Data Base is 1901–2000. Methods on compiling high resolution CRU TS 2.0 Global Data Base on Climate is described in Mitchell and Jones (2005). Example location of network of grids at the first stage of using our Data Base on regional climate one may see on Figure 3.2. The network there overlaid the same contours of river basins which had been shown on Figure 3.1.

Preliminary application of regional Data Base for hydrological and climatological purposes showed rather acceptable results and good compatibility with data on regular meteorological network.

Scientific background which was used to get data in the Table 3.1 consists of several independent components.

1. Adjusting of glaciers area values to the certain unified term. It was done by simple linear interpolation or extrapolation when we had at least two estimations of area.
2. Determination glaciers area $\boldsymbol{F}$ outside of known empirical temporal range. Firstly it could be done by linear extrapolation and secondly by means of equations:

$$F_{t+1} = F_t + \frac{dF}{dt}\Delta T \tag{1}$$

and

$$\frac{dF}{dt} = f(I_{Ac}, I_{Ab}) \tag{2}$$

or

$$\frac{dF}{dt} = f(\bar{T}_S) \tag{3}$$

where ΔT is time interval, I_{Ac} and I_{Ab} are indexes of yearly accumulation and ablation. Instead of I_{Ac} was used sum of precipitation for characteristic season and instead of I_{Ab} – mean summer air temperature$\bar{T}_S$. More detail information on getting and using equations (1–3) is described in Konovalov and Williams (2005) and Agaltseva and Konovalov (2005).

Table 3.1. Long term change of glaciers area in the upstream of Amudarya River (km^2).

Basins	Glaciers area, km^2				dF, km^2						dF %
	1961	1980	1991	2000	1961–1980	1980–1991	1991–2000	1961–1991	1980–2000	1962–2000	1961–2000
Vakhsh											
Obihingou	787.4*	705.1**	659.4***	608.5▲	82.3	45.7	50.9	128.0	96.6	178.9	**22.7**
Kyzylsu West	535.5*	486.4**	468.2***	449.7▲	49.1	18.2	18.5	67.3	36.7	85.8	**16.0**
Muksu	2061.7*	1987.5**	1946.3***	1887.1▲	74.2	41.2	59.2	115.4	100.4	174.6	**8.5**
Surhob RT.	394.4*	358.8**	339.1***	297.5***	35.6	19.8	41.6	55.4	61.3	96.9	**24.6**
Sum Vakhsh	**3778.9**	**3537.8**	**3412.9**	**3242.8**	**241.1**	**124.9**	**170.1**	**366.0**	**295.0**	**536.1**	**14.2**
Panj RT											
Yazgulem	337.9*	262.7**	221.0***	200.3▲	75.2	41.8	20.7	117.0	62.4	137.6	**40.7**
Vanch	333.2*	291.6**	268.4***	255.1▲	41.6	23.1	13.3	64.8	36.5	78.1	**23.4**
Gunt	514.6*	441.1**	400.2***	360.3***	73.5	40.9	39.9	114.4	80.8	154.3	**30.0**
Shahdara	205.5*	166.7**	145.2***	136.2***	38.8	21.5	9.0	60.3	30.5	69.3	**33.7**
Bartang	1701.7*	1382.1**	1204.5***	1146.0***	319.6	177.6	58.5	497.2	236.1	555.7	**32.7**
Panj (1)	51.2*	48.1**	46.4***	39.3***	3.1	1.7	7.1	4.8	8.8	11.9	**23.2**
Panj (2)	45.0*	44.0**	43.0***	40.9▲	1.0	1.0	2.1	2.0	3.1	4.1	**9.1**
Panj (3)	358.8*	268.9**	219.0***	210.6▲	89.9	49.9	8.4	139.8	58.3	148.2	**41.3**
Sum Panj RT	**3547.9**	**2905.2**	**2547.7**	**2388.7**	**642.7**	**357.5**	**159.0**	**1000.2**	**516.5**	**1159.2**	**32.7**
Panj LT											
Pamir R.	350∇			335.5▲							
Vagjir R.	278∇			269.3▲							
Vahandarya R.	757∇			651.3▲							
Panj R. (4)	1710∇			1586.6▲							
Kokcha R.	838∇			569.7***							
Kunduz R.	337∇			229.1***							
Sum Panj LT	**4270**			**3641.5**						**628.5**	**14.7**
Panj in total	**7817.9**			**6030.2**						**1787.7**	**22.9**
Amudarya upstream	**11596.8**			**9273**						**2323.8**	**20**

Notes: Surhob RT – right tributaries of the Surhob River, Panj R. (1) – right tributaries of Panj below mouth of Vanch River, Panj R. (2) – right tributaries of Panj between mouths of Yazgulem, Bartang and Gunt Rivers, Panj R. (3) – right tributaries of Panj above mouths of Gunt River, F_{gl} – glaciers area, Panj R. (4) – left tributaries of Panj from confluence of Pamir and Vahandarya Ribers till the mouth of Kokcha River, * – data of the USSR Glaciers Inventory (1971–1978), ∇ – data of Ivan'kov (1970), ** – data of Schetinnikov (1997), *** – computed by Konovalov, using his methods (Konovalov and Williams, 2005), ▲ – determined by Konovalov after processing remote sensing images obtained from Landsat 7 (scanner ETM+) and Terra (scanner Aster) satellites, dF – change of glaciers area for the proper time interval.

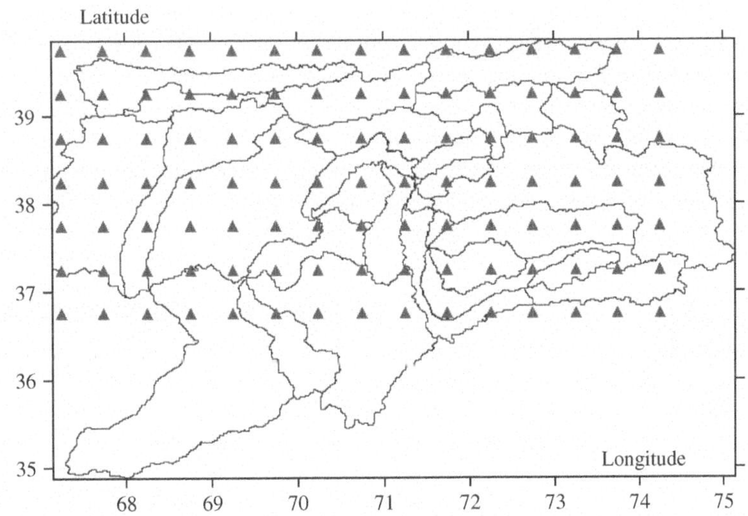

Figure 3.2. Location of meteorological network for Amudarya upstream.

3. Recognition and digitizing glacier contours on remote sensing images and processing sets of such contours by means of known GIS software, e.g. ENVI, ArcGis, IDRISI and others.

Since method 2 plays significant role for estimation of future glaciers area Table 3.2 presents some independent data on results of quality control for this method.

Table 3.2. Comparisons glaciers area determined by different methods.

Basin/Region	**Determinations F_{gl} in 2000 km^2**		**ΔF, %**
	$F_{gl} = f(\bar{T}_S)$	**Other estimations F_{gl} and their sources**	
		40.9 □	
Oigaing river (part) – TienShan	39.6	43.7 [Glazyrin and Schetinnikov (2001)]	−3.3
Pskem and Chatkal rivers – TienShan	119.9	107.8 [Glazyrin and Schetinnikov (2001)]	−10.4
Gissaro-Alai mountain region	1503.7	1579.0 [Glazyrin and Schetinnikov (2001)]	10.1
Pumku river (Tibet, China)	1356.5 (2001)	1330.0 (2001) [Tao Che *et al.* (2003)] □	−5.0
Obihingou river – Pamir	575.9	608.5 [Agaltseva and Konovalov (2005)] □	2.0
Kyzylsu West river – Pamir	397.3	449.7 [Agaltseva and Konovalov (2005,] □	−5.7
Yazgulem river – Pamir	214.6	200.3 [Agaltseva and Konovalov (2005)] □	−13.2
Vanch river – Pamir	238.2	255.1 [Agaltseva and Konovalov (2005),] □	6.7
			−7.1

Notes: 1. $F_{gl} = f(\bar{T}_S)$ – results obtained according to method in Agaltseva and Konovalov (2005), 2. symbol □ means, that glaciers area is determined by processing remote sensing images obtained from LANDSAT 7 and TERRA satellites. 3. ΔF is relative difference between the considered estimations.

As one may see the relative difference between glacier areas determined from remote sensing images and computed by dependence $F_{gl} = f(\bar{T}_S)$ varies from 2% till 13% that confirms reliability of the suggested method. Rather significant shrinkage glaciers area in the upstream of Amudarya river during 1961–2000 years corresponds well with estimations of glaciers fluctuations in the other mountain regions. Namely:

- As reported in Vilesov and Uvarov (2001) during of 1955–1990 area and volume of glaciers in Zailiiskiy Range (Kazakhstan) diminished on 29.2% and 32.3% correspondingly;
- By data Tao Che *et al.* (2003) total area of glaciers in the Pumku river basin (Tibet, China) was equal to 1556 km^2 in 1987 but in 2001 it become less on 14.5%;
- According to Kuzmichenok (2006) total glaciers area in Kyrgyzstan over 1950–1960 estimated as 8100 km^2, then over 1977–1980 it diminished till 7400 km^2 and by 2000 glaciers area again reduced till 6500 km^2. Thus, shrinkage of area during 40 years was equal to 19.8 %;
- The paper of Stokes *et al.* (2006) presents information on changes in the terminus position of 113 selected glaciers in the Caucasus between 1985 and 2000. The vast majority (~94%) of the glaciers have retreated since 1985, with a mean retreat distance of 121m (8.1ma–1).
- By using digitized glacier outlines inferred from the 1973 inventory and Landsat Thematic Mapper (TM) satellite data from 1985 to 1999, it was revealed in Paul *et al.* (2004) that area reduction of about 930 Alpine glaciers for the period 1985 to 1999 equalled to 18%.
- Shrinkage of glaciers area ΔFGL in Northern Tien-Shan for 1955–1999 (Bolch, 2006).

When the glaciated area decreases the area exposed to the sun also declines resulting is less melting water volumes and runoff from glaciers in spring and summer periods. However this effect is compensated by an increase in the air temperature. Therefore, it can be assumed that although an annual increase of the Greenhouse Effect within the next 20 to 30 years, considerable changes in the water resources are not expected. So probably changes in runoff for this time period will be within the limits of natural changeability, ranging from +3 to −7%. In the near future occurring reductions in water resources could be driven by a degraded glaciated area.

3.2 SNOW COVER AND HYDROLOGICAL REGIME IN THE MOUNTAINS OF CENTRAL ASIA

3.2.1 Tools and method

Six benchmark basins with different and complementary behaviours were chosen for the investigation (Table 3.3). Their areas vary from 1,890 to 7,580 km̀ (Figure 3.3) and for all basins, the runoff was monitored until the early 90's. For some of them the monitoring restarted after 2000.

Table 3.3. Main characteristics of the 6 benchmark basins (Figure 3.3).

Basin	River	Gauging station	Area (km^2)	% of glaciers (2000)
Vakhsh	Kyzylsu	Dombrachi	7 580	6%
	Muksu	Davsear	6 800	29%
	Obighingou	Sangvor	1 890	22%
Pyandj	Vanch	Vanch	2 050	13%
	Yazgulem	Yazgulem	1 960	10%
	Kudara	Kudara	4 120	14%

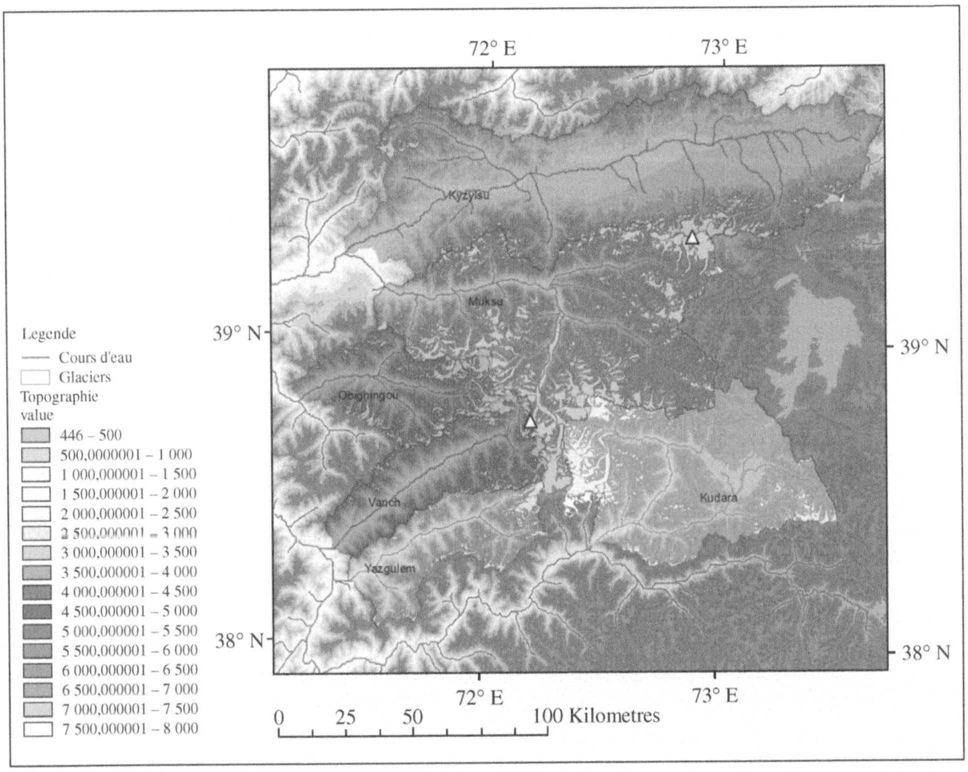

Figure 3.3. The 6 benchmark basins: (i) tributaries of the Vakhsh River, Kyzylsu at Dombrachi, Muksu at Davsear and Obighingou at Sangvor and (ii) tributaries of the Pyandj River, Vanch at Vanch, Yazgulem at Motravy and Kudara at Kudara. The white triangles locate the two main summits: in the centre of the map, the Ismael Somoni Peak (7,495m, former Peak of Communism), and in the north-east, the Ibn Sina Peak (7,134m, former Lenin Peak).

Three types of spatial information issuing from satellites sensors are used for the study:

- The digital elevation model at a 90m resolution (3" of arc) given by the *Shuttle Radar Topography Mission,* realized in February 2000 (http://www2.jpl.nasa.gov/srtm/).
- The climate data given by the reanalysis NCEP-NCAR at the nodes of a 2.5 degrees grid, since 1948 (http://www.cdc.noaa.gov/data/reanalysis/reanalysis.shtml).
- The Normalized Difference Index of Snow issuing of the information process at a 500m- resolution from the MODIS sensor (*Moderate Resolution Imaging Spectroradiometer*) on the TERRA satellite (http://modis.gsfc.nasa.gov/).

Digital elevation model SRTM

The raw data given by NASA present numerous lacks and sinks, which disturb the hydrological treatments. In order to limit the inconveniences, we used the retreated data for the *Consultative Group on International Agricultural Research* (Reuter *et al.*, 2007; Jarvis *et al.*, 2008).

Reanalysis climate data

Data resulting from climate models (Kistler *et al.*, 2001) assimilate satellite information and observed data. They give regional values, which can be used as indexes when local information are lacking. Past research shows that the temperature values given by the reanalysis in Central Asia are very likely doubtful. The main reason seems the strong influence of the relief, generating a great local variability, on the one hand, and the low accuracy of the observed data used for the calibration, on the other hand. Nevertheless, those data were especially used for the interval of 1990/2000, when the field observations are completely missing in the study area. Furthermore, the value of *precipitable water,* for the same reasons of lack of observed precipitation data, were carefully used.

Snow index given by the MODIS sensor

For this study, the *Normalized Difference Snow Index* (NDSI) of 48 images MODIS between May 2000 and May 2002 was downloaded from the data server http://nsidc.org/data/snowi. It consists of daily data with an acceptable limited cloud cover on the whole study area. We will see further in detail that the cloud cover impact is varying locally. An 8 days synthesis is also available, correcting sometimes the clouds inconvenient, but it has not been used here. The NDSI product of the MODIS sensor presents some weaknesses, especially due to shade effects, which can be corrected with a quite sophisticated method (Lopez *et al.*, 2008; Sirguey *et al.*, 2008), which has also not been used here.

3.2.2 Results and discussion

Figure 3.4 presents the hypsometric curves of the basins obtained after an extraction of the watersheds and the process of the SRTM DEM information. Those curves permit to organize the basins in 3 categories, taking also into account the climate characteristics: (a) Kyzylsu and Kudara Rivers basins with a relief of broad plains and elevated plateaux surrounded by the mountain ranges; the climate is quite dry (less than 400 mm of annual average precipitation). (b) Muksu and Obighingou Rivers basins with a relief of narrow valleys and more than 20 % of glaciered area; the annual precipitation average is above 1,000 mm. (c) The Vanch and Yazgulem Rivers basins with a fish bones shape at a globally lower elevation than the previous basins and an annual average precipitation around 800 mm.

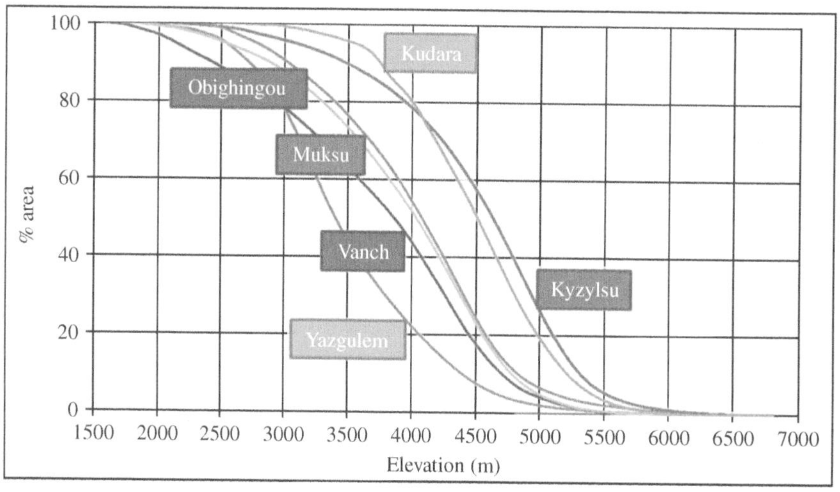

Figure 3.4. Hypsometric curves of the 6 benchmark basins given by the SRTM DEM (corrected data).

Figure 3.5 and Figure 3.6 present examples of results obtained with the MODIS snow product. On Figure 3.5, corresponding to a date of a maximum snow cover extension (caution! the snow cover is in grey colour in order to be visible in transparency with the basins limits), one observes that some glaciered areas do not appear snowy, especially close to both main summits (see Figure 3.3). At this step, we do not understand this anomaly, which not seems to be linked with the cloud cover, but more likely to steep rocks slopes without light reflection.

On Figure 3.6, the points of the interval May 2000–May 2002 are not linked with a curve because the snow cover is not accurately known between two dates of satellite information. Comparing the obtained results for the Obighingou and Kudara Rivers, one observes that for the first one, with steeper relief and stronger precipitation as the second one, the decreasing of the snow cover presents a lag of at least two months, on the one hand, and that the percentage of the area concerned by the maximum snow cover is significantly higher (90% vs 70%), on the other hand. This difference is easily justified by the different climate regimes. The dryer Kudara River basin is protected from the Westerlies by the orographic effect, despite the clouds cover, which seems more present, in reference with the lower useful number of MODIS scenes in this basin.

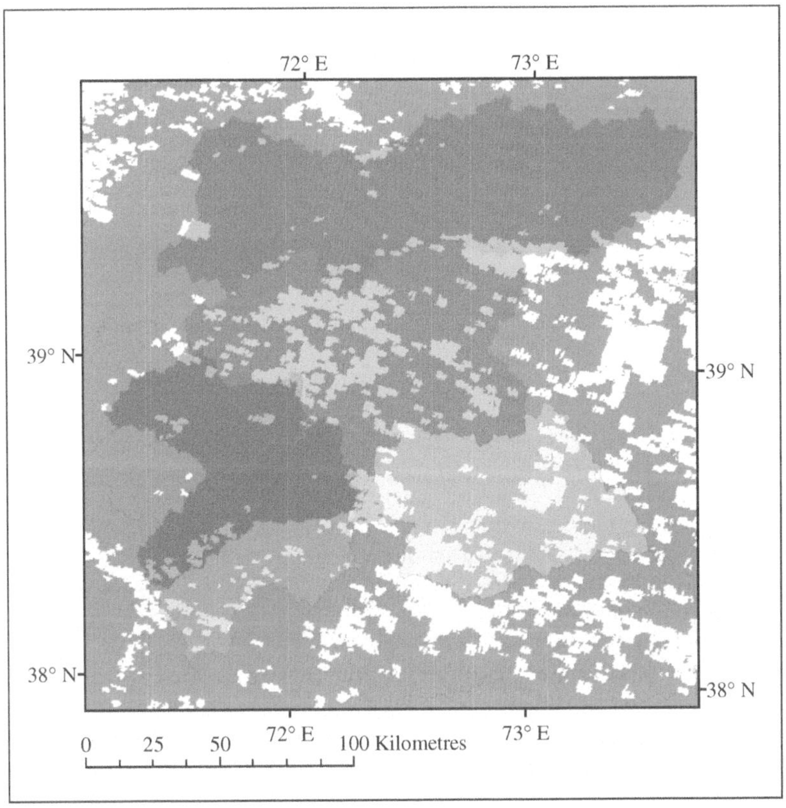

Figure 3.5 Snow cover obtained with MODIS snow product on 22th December 2000. The snow pixels are in grey color. The basins colors are the same as in Figure 3.3.

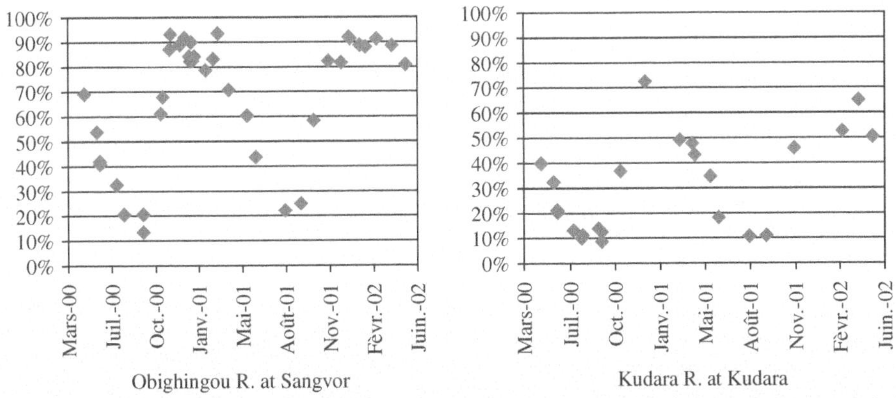

Figure 3.6. Dynamics of the snow cover in the Obighingou and Kudara Rivers basins obtained through the analysis of the MODIS snow product in 48 scenes between May 2000 and May 2002. We retained only the scenes where the percentage of clouds is less than 20% of the snow cover area. That corresponds to 38 points for Obighingou R. basin and 23 points for Kudara R. basin).

Applied to the Obighingou River basin, Figure 3.7 compares in standardized values the dynamics of the snow cover in the interval May 2000–May 2002 with the observed flow discharges at Sangvor and Yozgand (gauging station installed in 2002, downstream from Sangvor; its basin has an area of 5,651 km^2). Due to the absence of observed climate data, reanalysis values were introduced: temperature at the 600 hPa level (equivalent to an elevation of 4,200 m) and precipitable water, giving a trend for the true precipitations. One notes especially: (a) the duration of the snow cover permanence not seems to depend of the precipitation volume, while the water flow volume seems better correlated; (b) the low water discharge is remaining during the complete period of significant snow cover; (c) the snow cover is not generated by the main precipitations, which occur in summer; (d) the water discharges begin to increase before the snowmelt; it could be interpreted as a precocious contribution of melting of the glaciers or of the avalanches reaching lower altitudes.

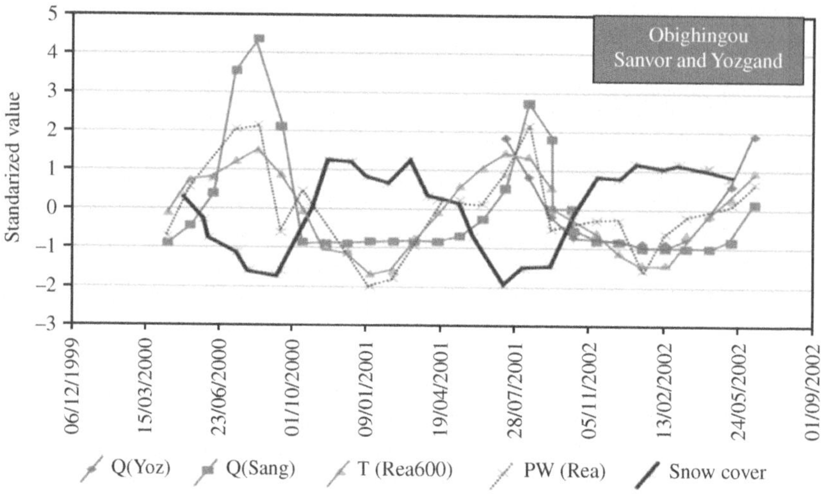

Figure 3.7. Snow cover and hydrologic regime: example of the Obighingou River basin. All the variables are presented as standardized values. Q(Yoz) and Q(Sang) are the flow discharges observed at the gauging stations of Sangvor and Yozgand. T(Rea600) is the temperature given by the reanalysis at the level 600hPa and PW is the value of the precipitable water (node: 40.0°N, 72.5°E). Snow cover corresponds to the MODIS snow product for the Obighingou River basin at Sangvor.

The Figure 3.8 presents for the 6 benchmark basins the distribution of the snow cover at 2 dates in December 2000 and May 2001. One notes in the Kudara River basin the anomaly commented for the Figure 3.5. We actually do not have any explanation for the absence of snow above an altitude 4,500 m in December 2000. Those graphs show clearly that the decreasing of the snow cover at the end of the winter season is concentrated in the highest altitudes (above 3,000 m). This statement is of high importance.

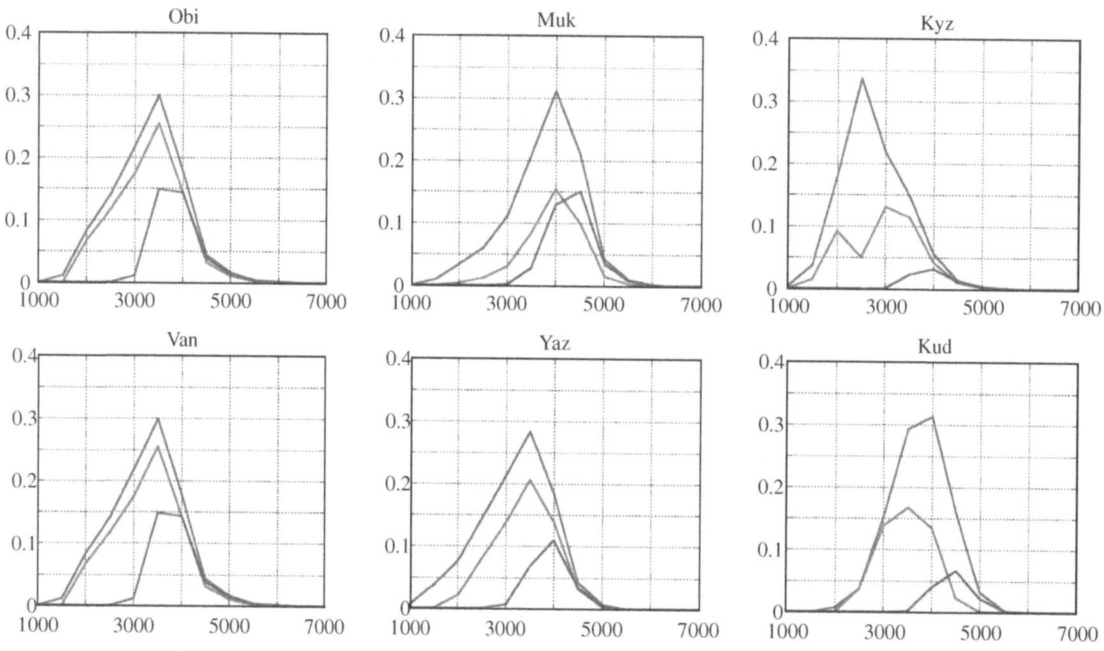

Figure 3.8. Dynamics of the snow cover regarding the altitude for the 6 benchmark basins. The upper curve of each graph corresponds to the hypsometric profile of the basin; the middle and lower curves correspond to the snow cover on 11th December 2000, respectively 15th May 2001.

Chapter 2 of this issue shows that, based on several SRES scenarios and all the main known Global Circulation Models, the increasing of the average temperature in the Pamir-Alay region could reach 5 to 6°C at the end of the Century. Taking as hypothesis that the mean elevation gradient of temperature of 6.5°C/1000m (Francou and Vincent, 2007) is reliable for this region, that would represent a right- translation of two intervals of the snow curves in Figure 3.8. As a consequence, the snow volume would decrease significantly because the basins area available for the snow accumulation is rapidly decreasing with the altitude, as shown by the hypsometric profiles of each basin.

3.3 EVALUATION THE CLIMATE CHANGES IMPACT TO THE AMU DARYA RIVER RUNOFF

The expected climate change impact on the upstream river regimes in the Amu Darya Basin was evaluated with help of reliable mathematical models for the runoff formation (described in more detail in Jayhun D.3.2 and Jayhun D.5.1). The applied mathematical model NIGMI describes a complete cycle of the runoff formation (Figure 3.9). It considers the main factors and processes such as precipitation, dynamics of a snow cover, evaporation, contribution of melting and rain water to the catchments, glacial runoff, runoff transformation and losses in basin. The model complex (set of models) consists of the model of snow cover formation for mountains basin, model of glacial runoff, transformation model for the runoff contribution from snowmelt and rain.

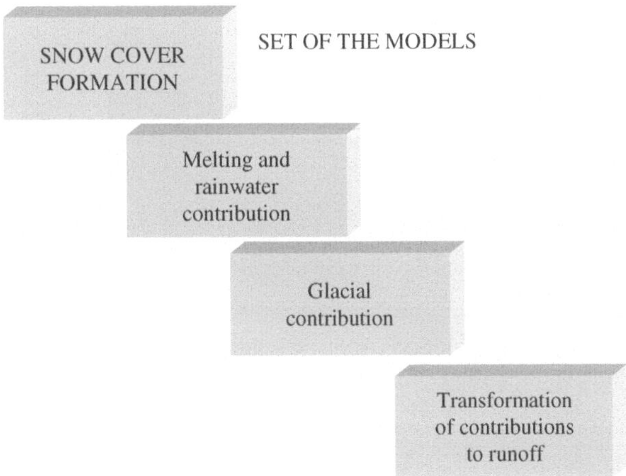

Figure 3.9 Model concept for the runoff formation.

Climate change impacts on the river runoff varying by the choice of applied climate change scenario and depends on the resulting precipitation rates. Due to the high natural variation in measured precipitation and some climate scenarios uncertainty (see Chapter 2) evaluative calculations were completed for three climate scenarios B1, A1B and A1F.

The results of mathematical runoff modelling for the runoff formation zones of selected river basins are given in Table 3.4 and Table 3.5.

Table 3.4. Runoff evaluations for vegetative period, from April to September, in respect to the air temperature and precipitation scenarios (multi-models approach, B1, A1B, A1F scenarios of greenhouse gases emission).

River, station	Q_{veg} m^3/s	Q_{veg} (per cent of basic norm) 2030			Q_{veg} (per cent of basic norm) 2050		
		B1	A1B	A1F	B1	A1B	A1F
Vakhsh-Komsomolobad	988	92	90	89	86	83	85
Kafirnigan-Tartki	255	95	96	105	89	87	99
Varzob-Dagana	78	89	86	94	85	80	88
Kyzilsu (south)-Samanchi	97	86	84	89	83	80	83
Amudarya-Kerki	3022	90	88	89	83	79	84

Table 3.5 Runoff evaluations for October-September time period in respect to the air temperature and precipitation scenarios (multi-models approach, B1, A1B, A1F scenarios of greenhouse gases emission).

River, station	$Q_{Oct\text{-}March}$ m^3/s	$Q_{Oct\text{-}March}$ (per cent of basic norm) 2030			$Q_{Oct\text{-}March}$ (per cent of basic norm) 2050		
		B1	A1B	A1F	B1	A1B	A1F
Vakhsh-Komsomolobad	602	97	96	95	93	90	92
Kafirnigan-Tartki	167	112	111	120	108	107	118
Varzob-Dagana	55	91	88	94	89	86	92
Kyzilsu (south)-Samanchi	78	96	93	98	94	92	96
Amudarya-Kerki	1960	98	97	98	94	92	96

Figure 3.10 is given as examples of climate change impacts on the river runoff for Vakhsh and Kyzylsu. The river runoff assessment identified for the climate changes scenarios that:

- if climate scenarios on precipitation & air temperature changes (scenario A1B) are realised for the Vakhsh, Kafirnigan, Kyzylsu and other river basins until 2030, a decreasing runoff trend will occur.
- air temperature long term changes (2050) scenario A1B leads probably to runoff decrease in the Vakhsh, Kyzylsu and other river basins. The runoff changes are expected to be in a range of 10–12%;
- the situation like above is expected in Amudarya river basin for scenarios B1 and A1F by 2050 (Figure 3.11 and 3.12).

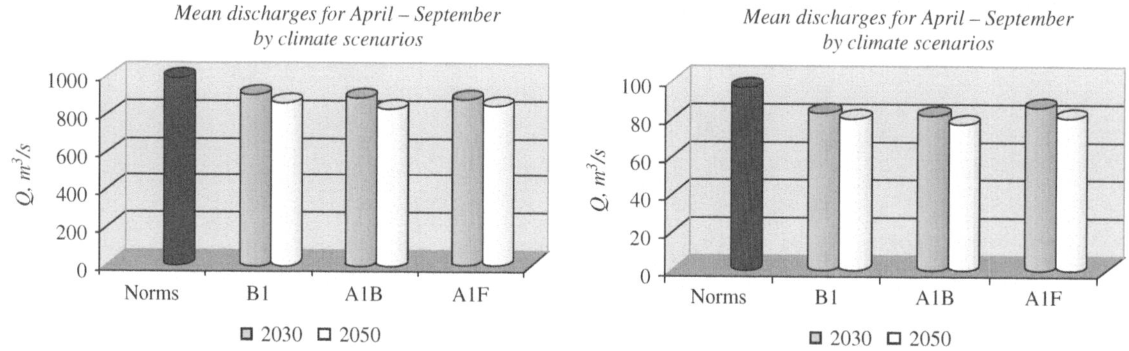

Figure 3.10 Mean discharges for vegetative time period (April – September) calculated per air temperature and precipitation scenarios for Komsomolobad water gauge station, Vakhsh river basin (left figure) and for Kyzylsu river- Samanchi gauge (right figure).

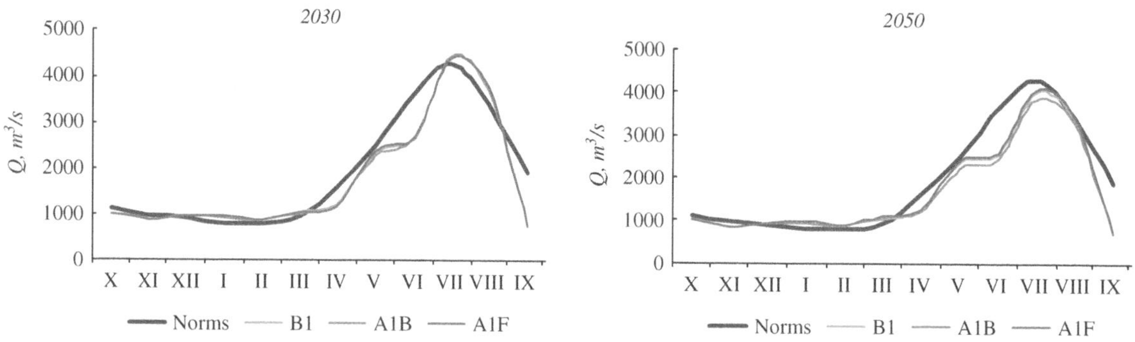

Figure 3.11. The runoff hydrographs for Amudarya river – Kerki gauge calculated per air temperature and precipitation scenarios.

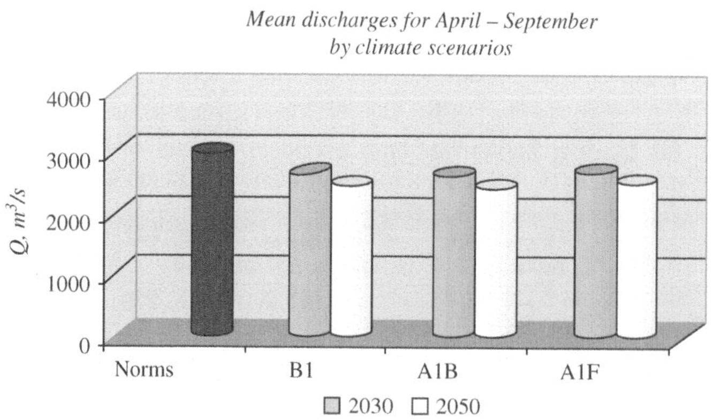

Figure 3.12. Mean discharges for vegetative time period (April – September) for Amudarya river- Kerki gauge calculated per air temperature and precipitation scenarios.

Different climate change scenarios were simulated using mathematical runoff models for mountainous rivers. The results of scenario A1B show the trend toward a river runoff decrease caused by changes in air temperature and precipitation. For the long term perspective (by 2050) such decrease will range between 10 to 12%, taking current conditions as basis. Considering present water deficits this reduction will become a serious problem for the countries in the middle and down stream part of the Amudarya. Moreover, changes in the hydrograph are expected to be negatively affecting the agricultural sector.

On the other hand, analyses of hydro-meteorological data could lead to the conclusion that under an increase in precipitation between 5 and 15% until 2050 and 10 to 25% until 2080 will occur the runoff could be formed in the present magnitude.

The described method of runoff evaluation holds a certain uncertainty with relation to the natural variability of precipitation. Nevertheless it could be supposed that for short term perspective no increase in the runoff can be expected under the existing precipitation pattern even if some precipitation rates will rise.

Serious reductions in water resources are expected to the middle of the 21st century. At the same time the implementation of adaptation measures to climate change as result of both human activities and natural factors has already turned out as one of the most important problems.

The results clearly demonstrate the need to strengthen and deepen the knowledge on climate change impacts and the implications for the general water resources availability. Regional water resources monitoring systems as part of an integrated water resources assessment could help to reduce arising and intensified problems.

In this context, it should be taken into consideration that the presented model results give only an idea on changes in river runoff under the defined future climate scenarios.

3.3.1 Changes in the river runoff based on the simulation of an artificial time series for precipitation and temperature

The other approach to the assessment of the future Amudarya river runoff based on the use of the simulated time series of precipitation and air temperature for the period from 2000 to 2100.

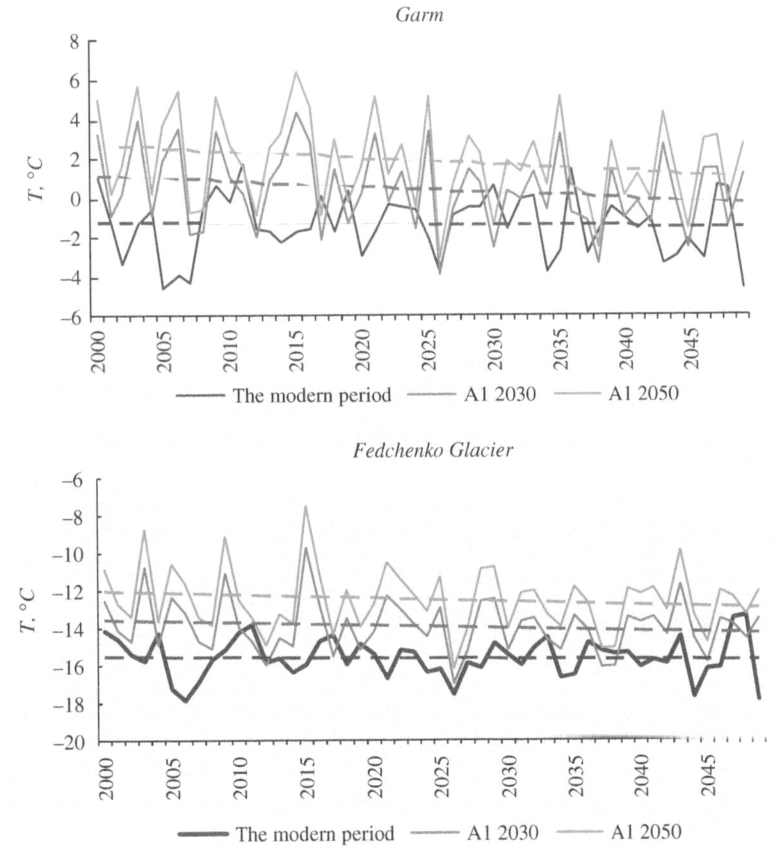

Figure 3.13. The interannual fluctuations of the air temperature averaged (January–March) by data for Garm and Fedchenko glacier meteorological stations.

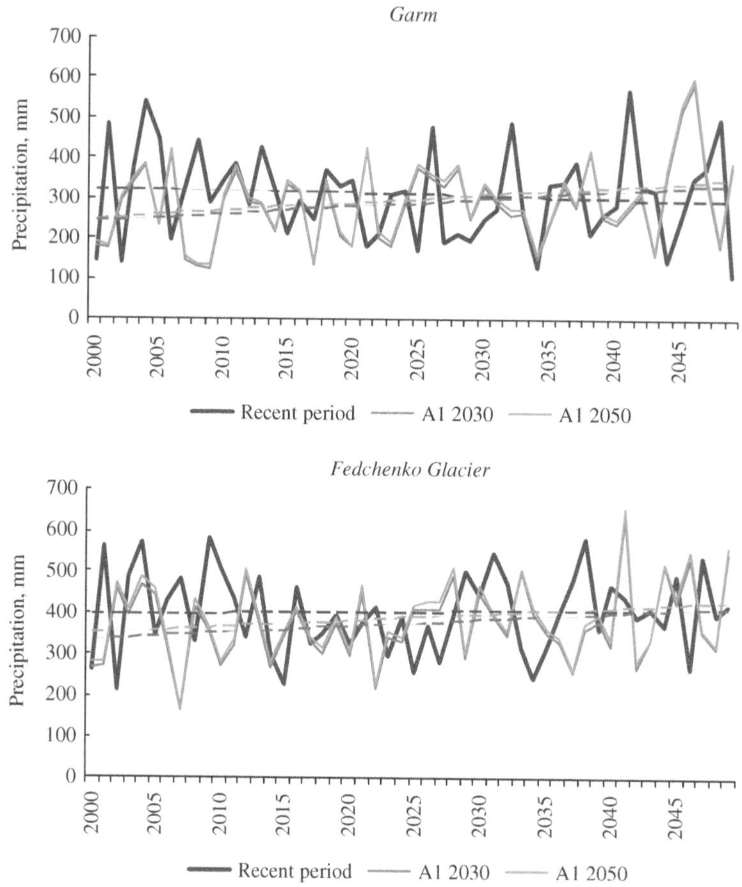

Figure 3.14. The interannual fluctuations of the seasonal (January–March) sums of precipitation measured for Garm and Fedchenko glacier meteorological stations.

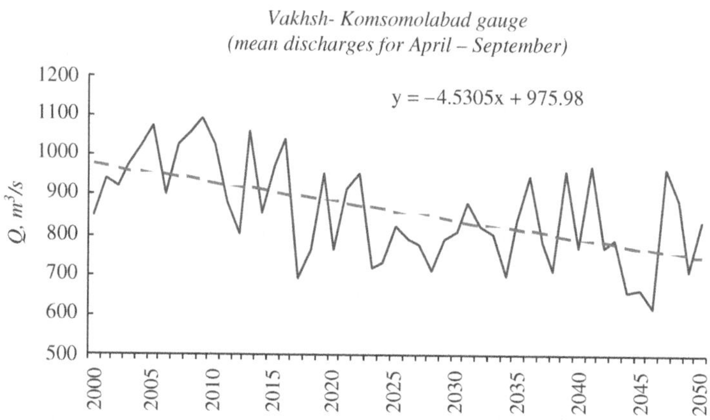

Figure 3.15 The expected changes in vegetative and mean annual runoff through the 2000 – 2050 time period for Vakhsh – Komsomolabad (based on present climate conditions).

Long time series of runoff (inflow) are necessary for simulating water economic estimations in Amu-Daria River basin. The length of inflow time series should be sufficient for the required reliability of such economic estimations.

Inflow time series should not covering a period of several decades because of the difficulties during the calculation process especially under the conditions of the Upper Amu-Darya River basin. For this, only fragmentary observation data (10–15 years) is available, in particular for the Afghan part of catchment area. By this, it is necessary to use algorithms for completing and setting-up artificial runoff time series.

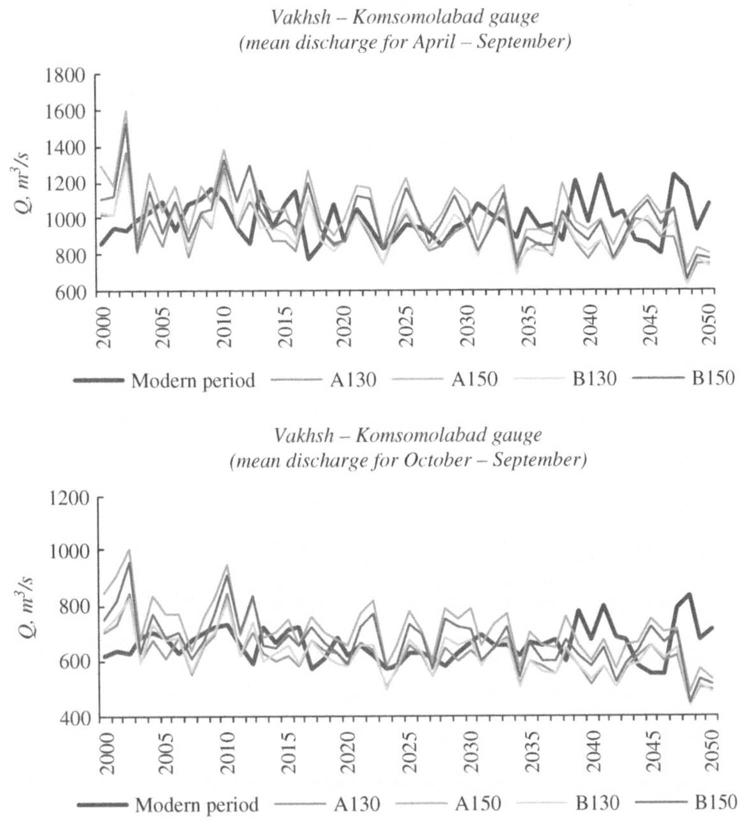

Figure 3.16. Comparison of simulated runoff by different artificial time series of temperature and precipitation for the Vakhsh – Komsomolabad river basin.

Monthly measurement data of 24 stations (precipitation) and 17 stations (temperature) were used for the calculation of the runoff time series using the NIGMI model[1] and pre-defined 4 climate scenarios for 2030 and 2050 to reference points of Amudarya river basin.

The generated artificial monthly time series assuming 4 climate scenarios were applied for the runoff calculation in the Pjanj and Vakhsh rivers basins. The calculations for the future runoff time series were performed considering four variants:

- for the 2030 climate change scenario and realization of the SRES A1F1 emission scenario;
- for the 2050 climate change scenario and realization of the SRES A1F1 emission scenario;
- for the 2030 climate change scenario and realization of the SRES B1 emission scenario;
- for the 2050 climate change scenario and realization of the SRES B1 emission scenario.

The statistical characteristics of the scenarios developed in chapter 2 were applied for the simulation of the monthly meteorological time series. Moreover, the simulation of the meteorological data was also done for the current climate conditions, without taking climate change into account.

Figures 3.13 and 3.14 show the simulated values of precipitation and air temperature averaged for different time periods at some stations in the Amudarya runoff formation area, for current conditions as well as for the expected climate change conditions for 2030 and 2050.

The analysis of the simulation results show a high natural variability in the precipitation and air temperature and different trends for the observation points. An upward trend in precipitation is observed during winter and spring time and at some stations the trends of air temperature differ for the seasons. Consequently, the calculations of the future runoff time series were performed using the same four variants as mentioned.

In addition, scenarios were simulated calculating runoff time series with present climatic conditions also valid for the future, i.e. without climate change impacts.

[1] More explanation on the NIGMI modeling procedure can be given by the Hydrometeorological Research Institute of Uzbekistan

The results of the calculated inflow time series for the Nurek reservoir (at the Vakhsh river) using available meteorological data are given in the Figures 3.15 and 3.16. There is a clearly defined tendency of runoff reduction, especially during the vegetation period (April – September) when the need for water resources is especially high.

The same calculation pattern using the regional hydrological model (NIGMI's model) were implemented also for other rivers in the runoff formation area of the Amudarya river.

The results are shown in Figures 3.17 and 3.18. A slightly different situation results in the Kafirnigan river basin. Under given conditions the calculated run-off data are ranged within the observed data or even with a rising tendency. This can be explained by the fact that the basin is more favourable to get wet air masses and an increasing precipitation under scenario conditions.

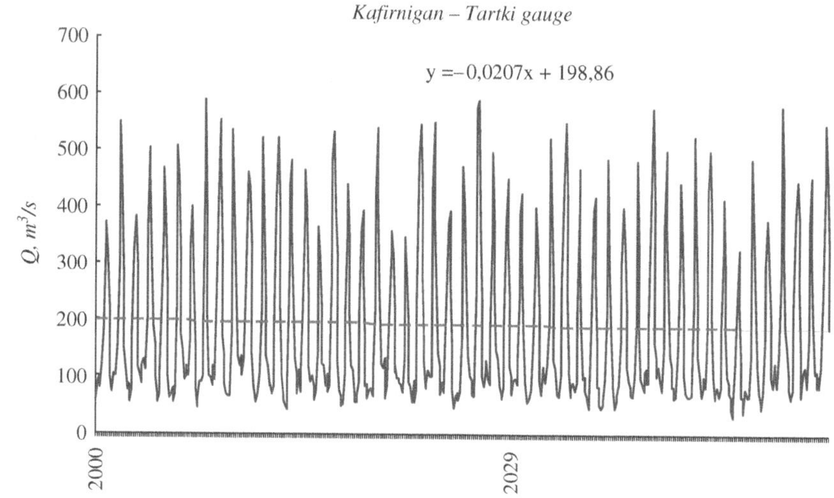

Figure 3.17 The expected changes in monthly runoff through the 2000–2050 time period for Kafirnigan – Tartki (based on present climate conditions).

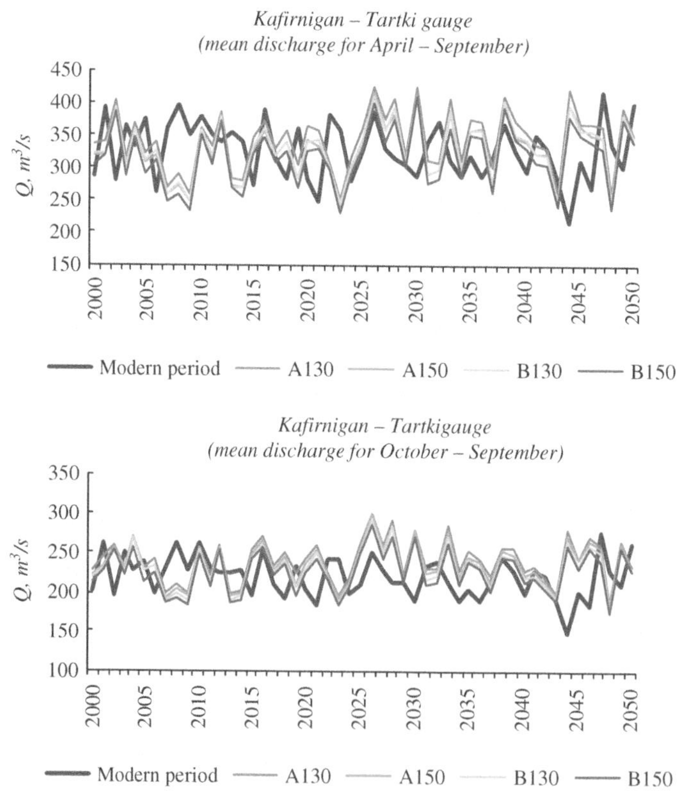

Figure 3.18 Comparison of simulated runoff by different artificial time series of temperature and precipitation for the Kafirnigan river basin.

Integrated assessment of the Amudarya river runoff for the Kerki water gauge station has revealed a general tendency to the runoff reduction for middle-term and long-term perspective. This is in line with the expected climate change in the region (Figures 3.19 and 3.20).

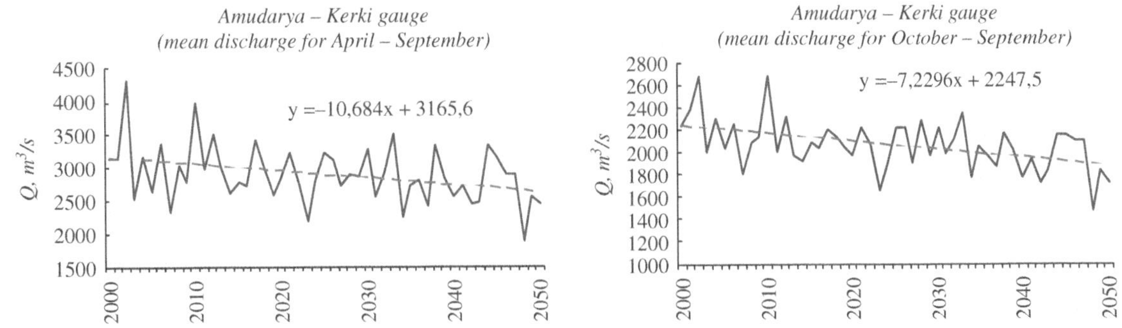

Figure 3.19 The expected changes in vegetative and mean annual runoff through the 200–2050 time period for Amudarya – Kerki (scenario A130).

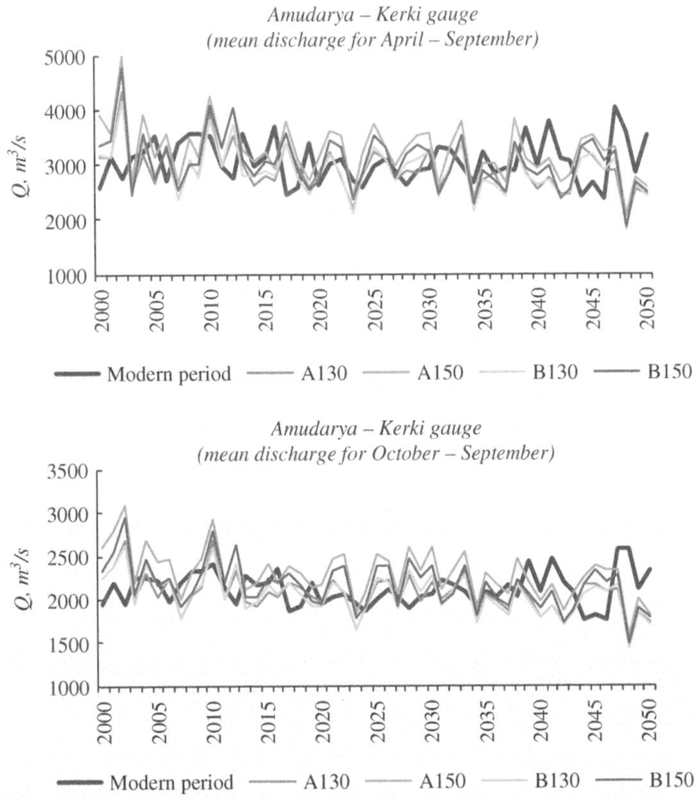

Figure 3.20 Comparison of simulated runoff by different artificial time series of temperature and precipitation for the Amudarya – Kerki basin.

Summarising the results of the runoff assessment for the period up to 2050 based on the simulated series of the meteorological data allow:

- to evaluate possible change in runoff for the period 2000–2050;
- to apply the series of the monthly runoff for the evaluation of hydro-economic and other analyses;
- to revise and re-evaluate the climate basis with new runoff estimations based on a new approach for designing change climate scenarios.

The presented results are useful to be used as basis for the development of new water management strategies on adaptation to climate change considering hydro-power.

3.4 REFERENCES

Agaltseva, N.A. and Konovalov, V.G. (2005) Projected change of glaciers size and river runoff in the different scenarios of future climate. *Exchange of intellectual property*, **4**(8), 37–47 (in Russian).

Bolch, T. (2006) GIS- und fernerkundungsgestützte Analyse und Visualisierung von Klima- und Gletscheränderungen im nördlichen Tien Shan (Kasachstan/Kyrgyzstan) mit einem Vergleich zur Bernina-Gruppe/Alpen. Als Dissertation genehmigt von den Naturwissenschaftlichen Fakultäten der Universität Erlangen-Nürnberg Tag der mündlichen Prüfung: 03.07.2006.

Francou, B. and Vincent, C. (2007) *Les glaciers à l'épreuve du climat*. IRD Editions, Belin, Paris.

Glazyrin, G.E. and Schetinnikov, A.S. (2001) Current and future state of Western Tien-Shan Glaciers. *Pros. SANIGMI*, **161**(242), 5–14 (in Russian).

Glazyrin, G.E. and Schetinnikov, A.S. (2001) State of Gissaro-Alai Glaciers in the last decades and its possible dynamics related to the future climate change. *MGI*, **90**, 126–129 (in Russian).

Inventory of USSR Glaciers (1971–1978) Leningrad, "Hydrometeoizdat" Publishing House, **14**(3), 7–12 (in Russian).

Ivan'kov, P.A. (1970) Contemporary glaciation in the upstream and lef tribitaties of Amudatya river. *Proc. VGO*, **102**(3), 239–244 (in Russian).

Jarvis, A., Reuter, H.I., Nelson, A. and Guevara, E. (2008) Hole-filled seamless SRTM data V4. International Centre for Tropical Agriculture (CIAT). Available from http://srtm.csi.cgiar.org

Jayhun (2009) D.3.2 Report on future changes in available water resources. JAYHUN Interstate Water Resources Risk Management. "Towards a Sustainable Future for the Aral Basin," Specific Targeted Research Project Contract No. 516761.

Jayhun (2009) D.5.1 Implementation of modelling tools. JAYHUN Interstate Water Resources Risk Management. "Towards a Sustainable Future for the Aral Basin," Specific Targeted Research Project Contract No. 516761.

Kistler, R., Kalnay, E., Collins, W., Saha, S., White, G., Woollen, J., Chelliah, M., Ebisuzaki, W., Kanamitsu, M., Kousky, V., Doul, H. v. d., Jenne, R. and Fiorino, M. (2001) The NCEP/NCAR 50-Year reanalysis. *Bulletin of the American Meteorological Society*.

Konovalov, V.G. and Desinov, L. (2007) Remote sensing monitoring of long term regime of the pamirs glaciation. Remote sensing for environmental monitoring and change detection (Proceedings of Symposium HS3007 at IUGG2007, Perugia July 2007). IAHS Publ. 316, 2007. In press.

Konovalov, V.G. and Williams, M.W. (2005) Long term fluctuations of glaciers and river runoff within Central Asia at the current climate conditions. *Meteorology and Hydrology*, **9**, 69–83 (in Russian).

Kuzmichenok, V.A. (2006) Monitoring of water, snow and glacial resources of Kyrgyzstan. Regional workshop on assessment of snow-glacier and water resources in Asia. Almaty, Kazakhstan. 28–39 November 2006, 125–143.

Lopez, P., Sirguey, P., Arnaud, Y., Pouyaud, B. and Chevallier, P. (2008) Snow cover monitoring of Northern Patagonia Icefield using MODIS satellite images (2000–2006). *Global & Planetary Change*, **61**, 103–116.

Mitchell, T.D. and Jones, P.D. (2005) An improved method of constructing a database of monthly climate observations and associated high-resolution grids. *Int. J. Climatol.*, **25**, 693–712.

Paul, F., Kääb, A., Maisch, M., Kellenberger, T., and Haeberli, W. (2004). Rapid disintegration of Alpine glaciers observed with satellite data. *Geophysical Research Letters*, **31**, L21402.

Reuter, H.I., Nelson, A. and Jarvis, A. (2007) An evaluation of void filling interpolation methods for SRTM data. *International Journal of Geographic Information Science*, **21**(9), 983–1008.

Schetinnikov, A.S. (1997) Morphology of glaciation inside of Pamiro-Alay river basins as related to 1980. Reference Book. Tashkent, "SANIGMI" Editorial Board (in Russian).

Sirguey, P., Mathieu, R., Arnaud, Y., Khan, M., Chanussot, J. (2008) Improving MODIS spatial resolution for snow mapping using wavelet fusion and ARSIS concept. *IEEE Geoscience and Remote Sensing Letters*, **5**(1), 78–82.

Stokes, C.R., Gurney, S.D., Shahgedanova, M., Popovnin, V. (2006) Late-20th-century changes in glacier extent in the Caucasus mountains, Russia. *Journal of Glaciology*, **52**(176), 1–11.

Tao Che, Xin Li, Li Zong Wu and Rui Jin (2003) Inventory of glaciers and glacial lakes and the identification of potential glacial lake outburst floods (GLOFs) Affected by global warming in the mountains of Himalayan region. http://www.jamstec.go.jp/jamstec-e/iugg/index.html

Vilesov, E.N. and Uvarov, E.N. (2001) Evolution of contemporary glaciation in Zailiiskiy Alatau Range in XX centenary. "Kazak Yniversity" Publ. House, Almaty, p. 252 (in Russian).

Chapter 4

A water mass balance for the irrigation massifs

Malika Ikramova, Oleg Lisenko, Anatoly Sorokin, Pierre Chevallier, David Rycroft, Bernard Pouyaud, Marie Mojaisky and Philippe Vauchel

4.1 EXISTING WATER SUPPLY TO NATIONS, IRRIGATION SCHEMES AND THE ENVIRONMENT

The identification of the real water demands is considered essential in addressing the water saving potential within the interstate water sharing discussions. Despite the fact that water losses account for more than 50% of demand, there remain claims for more water by the different riparian users. Providing sound information on real water demands after exploiting reasonable savings should strengthen the position in future discussions on the allocation of water resources.

The objective was to review existing data sources and data bases, as well as information systems and update these, adding new data received from investigations and analysis of the current situation, including data on Turkmenistan and Afghanistan. This study collated available hydrological information on water availability and consumption from existing data sources. The Jayhun database based on GIS visualisation was developed and all available data included. It covers the basin, states, irrigation projects and development trends; it provided input for the WEAP model (Chapter 5); and is application-oriented to the surface water and land resources, energy, environment, economic analysis.

4.1.1 Review of existing data sources and data bases of official water use statistics

At present many different types of data and information sources are available from different organisations: the BVO Amudarya, the Ministry of Agriculture and Water Resources of Uzbekistan (MAWR), each large irrigation system and the Scientific Information Centre (form ICWC).

In the mid of 90s national and regional databases were developed. Water and land resources management regional databases have also been developed by TASIS, UNDP, USAID funded projects (WARMAP, WUFMAS, RPBAM, EPIC). In addition GEF, IASF, BVO Amudarya, IC ICWC (WARMIS) have developed databases appropriate to their own activities. Uzbek companies have also developed databases within their sphere of activity such as: UZGIDROMET (surface water), State Statistical Committee (population, agriculture, food production), Geology Committee (ground water), MAWR (land and water resources distribution and consumption, food production), UZBEKENERGY (water energy). The scientific institutions have developed GIS databases such as: WP (analyses of water – salt balance of the Amudarya river 1978–2000), EPIC (Amudarya delta model), GAMS (by A. Savitskiy), ASBMS (Aral Sea basin hydro ecological monitoring digital system for 1980–2001).

 Edited by Oliver Olsson and Melanie Bauer. ISBN: 9781843393085. Published by IWA Publishing, London, UK.

All these data sources were not included in one reliable and convenient database. The several data sources were limited to the organisations and were not linked to other data sources or organisations. Therefore most data sources were viewed separately or were only partly interrelated. The data received from one source hardly corresponds to another source.

The developed JAYHUN database based on GIS visualisation (further JDB) online version installed in the Water Balance and Planning Department of the MAWR RU will be available to the basin management units and water users as Amu-Kaskadarya, Down-Amu, Amu-Bukhara Departments and Amudarya basin management unit (BVO) of the ICWC and etc. The database will be maintained and regularly updated by SANIIRI.

JDB was developed on a GIS platform. The basin map was activated in the ArcView program: the total basin area and the zones divided in to three sub-sections with more detailed introduction of each area: VDU – upper, SDU -middle and NDU – lower (Figure 4.1).

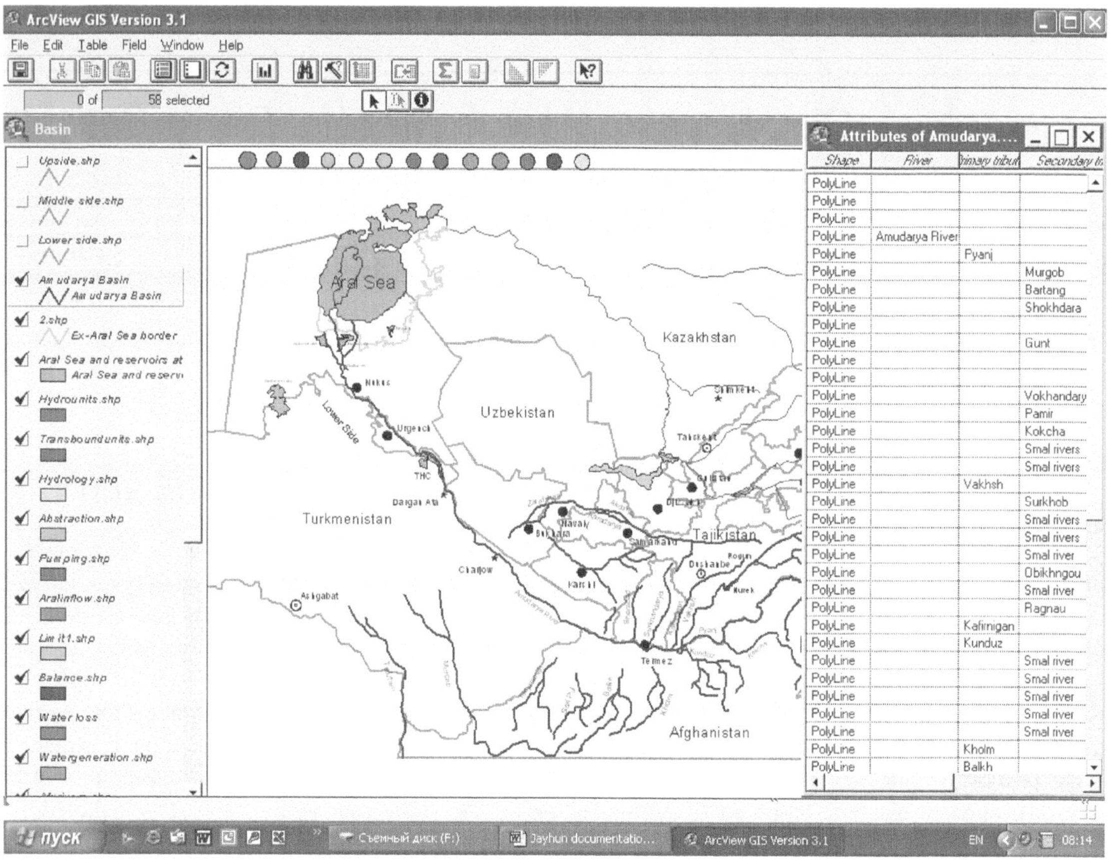

Figure 4.1. The Amu Darya Basin map implemented in the Jayhun DB and example data sheet.

The map objects were linked to Excel tables with data obtained from Uzgidromet, MAWR, Uzvodinspektsiya, UZVOD, THC management unit, BVO Amudarya, Surkhandarya irrigation systems management unit, Lower Amudarya Basin management unit and Uzkommunhizmat.

Data concerning quality and quantity of the water resources, technical characteristics of the water regulating bodies (hydro-technical units) assigned to generation, transfer, runoff control, consumption and drainage have been adjusted by the WP4 based on extended periods of investigation by observation, measurement and monitoring, experiments and by calculations.

The JayhunDB was developed as a tool to support risk management:

- it covers the basin, states, irrigation projects and their development trends,
- it is the basis for the WEAP model in WP5 (D.5.2),
- it is oriented to the surface water and land resources, energy, environment and economic analysis,
- it was created using EXCEL, ArcView 3.2, Surfer softwares.

There are no limits to the data storage and the Jayhun Database will be continuously updated. Additionally a detailed GIS map and data on water consumption by irrigation system was developed for the Amu-Kashkadarya, Amu-Zang

(Surkhandarya), Amu-Bukhara and Nijniy Amu-Darya basins. Furthermore the database was enhanced with data on the demand for ecosystems, including the Aral Sea and Priaralie.

Water resources formation in Afghanistan and irrigated areas including future irrigation projects development were included in the Database.

Amudarya hydrology for all stations, including Kelif, Kerki (Atamurat), Darganata, Tuyamuyun, Kipchak, Samanbay and Kiziljar were included. Figure 4.2 shows exemplary time series in the Jayhun DB for the period from 1989 to 2008 September. Due to the importance of the Termez gauging station as the officially accepted control post (by Uzbekistan and Turkmenistan) for the middle stream before the Amu Darya flows into Turkmenistan territory, the Termez station data were included for the period from 1999 to 2009.

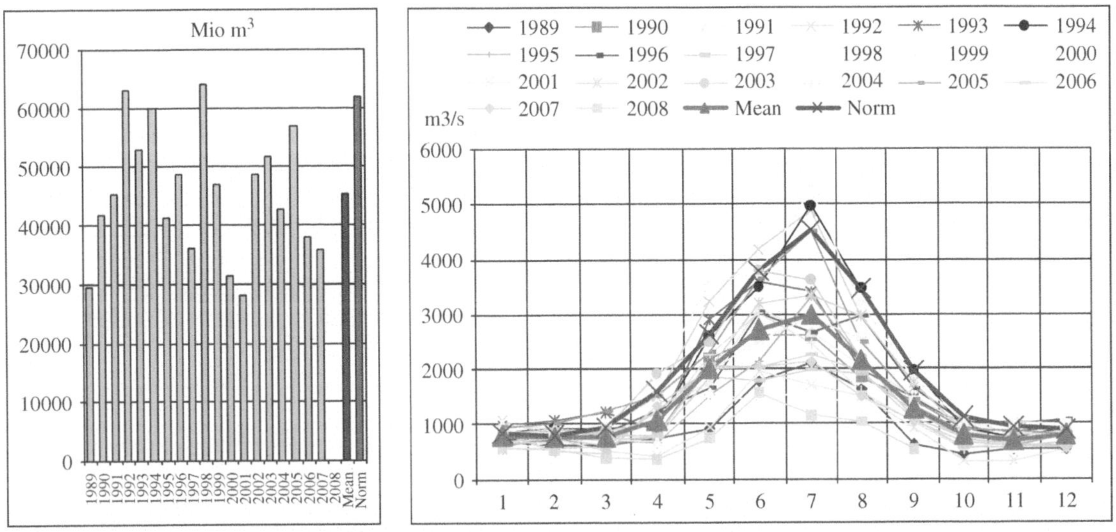

Figure 4.2. Water flow at Atamurat Station (Kerki) including the 2008 dry year.

The JAYHUN database and GIS map was enhanced by the inclusion of detailed water units within the irrigation projects in Aksu, Yakkabag-Guzar, Mirishkor, Karshi, Eski-Angar (Kashkadarya region) and Shakhrud-Dustlik, Amu-Karakul, Kharkhur-Duoba, Tashrabot-Jilvan, Tashrabot-Urtachul (Bukhara region) (Figure 4.3), Kizketken-Kegeyly (Karakalpakstan) in Uzbekistan. Depreciation of canals and regulators and other water units, drainage systems were analyzed to fix water delivery efficiency and water loss rates. As well as operation terms in different water conditions: deficit, normal and excess water.

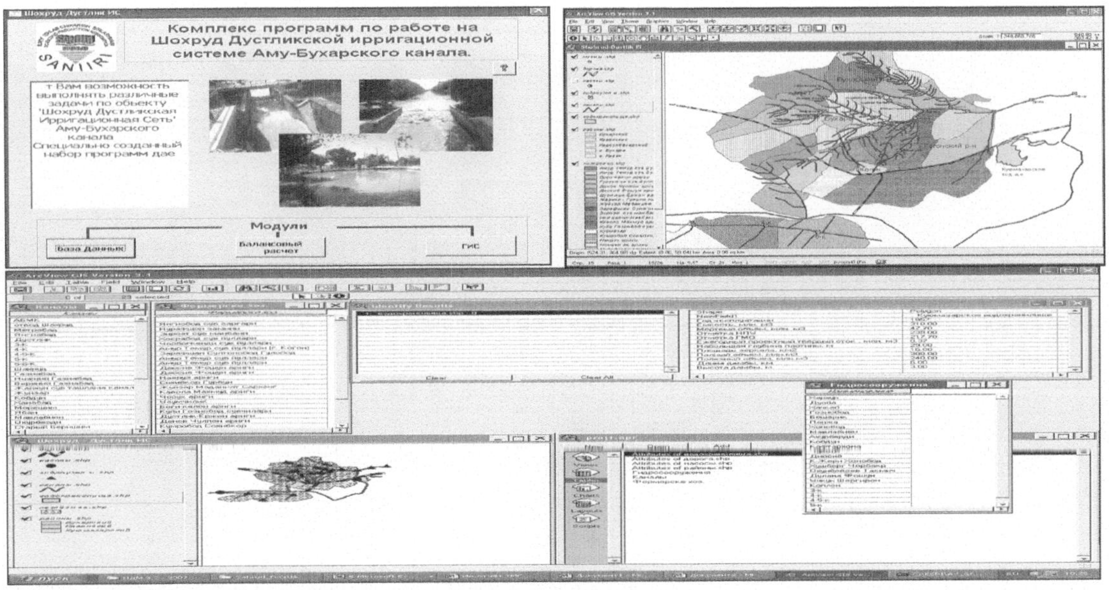

Figure 4.3 View of detailed database for the Bukhara region.

The WP4 developed a Database module "BIS" – basin information system software for Kashkadarya and Bukhara region, established in the basin management organizations and applied for daily operation. The "BIS" is composed of three modules: a GIS-map, an ACCESS Database for monitoring of water delivery and consumption and an EXCEL water balance calculation module. A training course for the developed program was organized for the basin management staff.

4.1.2 Specification of data base structure for risk analyses

In order to analyze the risks of water deficit a water balance scheme for water abstraction, consumption and return water was developed for 7 irrigation projects in three different water availability regions of Uzbekistan:

- Aksu, Yakkabag-Guzar, Mirishkor, Karshi, Eski-Angar irrigation systems in Kashkadarya region. The region takes 2200 Mio m^3 water directly from Amudarya through Karshi canal, 900 Mio m^3 water from Kashkadarya and 320 Mio m^3 water from Zarafshan (two tributaries of Amudarya do not reach the river at present),
- Shakhrud-Dustlik, Amu-Karakul, Kharkhur-Duoba, Tashrabot-Jilvan, Tashrabot-Urtachul irrigation systems in Bukhara region, taking water through Amu-Bukhara canal,
- Kizketken-Kegeyly irrigation system in Karakalpakstan., taking water through the Kizketken canal.

Balance calculations were undertaken to identify water distribution and establish a water deficit/or overflow, water losses in the irrigation systems. The water balance scheme was tested in order to find the optimal scenario for water distribution, to increase water use efficiency, to analyze and manage the irrigation systems. Recommendations to improve water supply were developed. The computational module for balance calculation includes a line scheme for the canals, includes its code, flow rate, water abstraction, and types of connection such as confluence or streamflow separation (Figure 4.4).

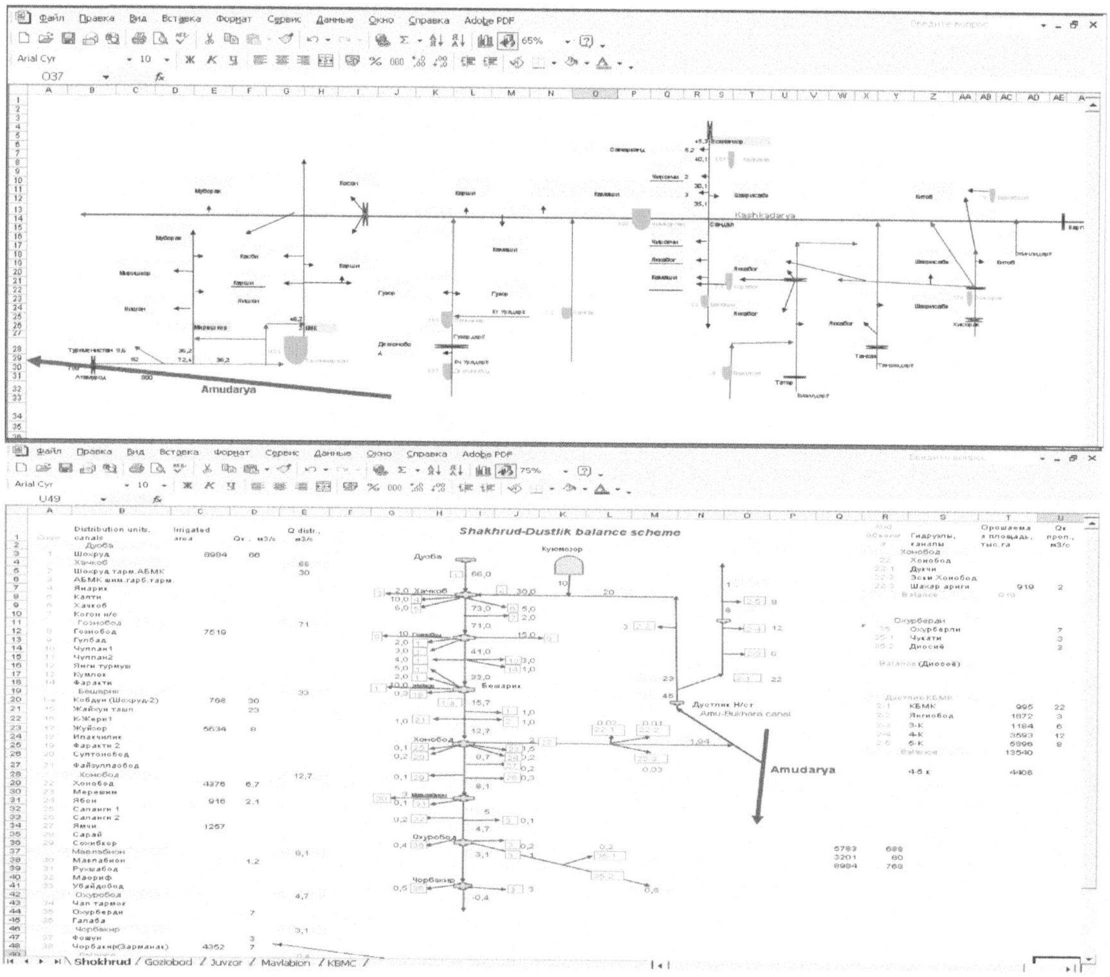

Figure 4.4. Water balance computing module for Amu-Kashkadarya irrigation project and for the Shakhrud-Dustlik irrigation system on the Amubukhara canal.

The following scenarios were tested: (i) For dry periods it was accepted as a criterion – minimum crop loss and maximum and accurate satisfaction of the water users demand. (ii) For high water periods it was based on prevention of flooding the territories, existence of additional capacity in the reservoirs, capacity of canals and drainage systems. Based on the test results an assessment of water distribution scenarios was carried out for:

- total water deficit in a system for time intervals,
- total limitation of water supply and deviations from demand,
- rate and duration of dry periods,
- total operating capacity of the existing reservoirs,
- total water volume accumulated/or outflowing in considered periods,
- outflow in the absence of regulating capacities,
- evaluation of water use efficiency.

Water abstraction to the Karshi and Amu-Bukhara canals was analyzed to identify the water availability of the irrigation projects in these areas (Figure 4.5 and Figure 4.6). The Figures show that water intake depends on the flow rate in the Amu Darya. A water supply plan was developed taking into account pumping from the Amu Darya.

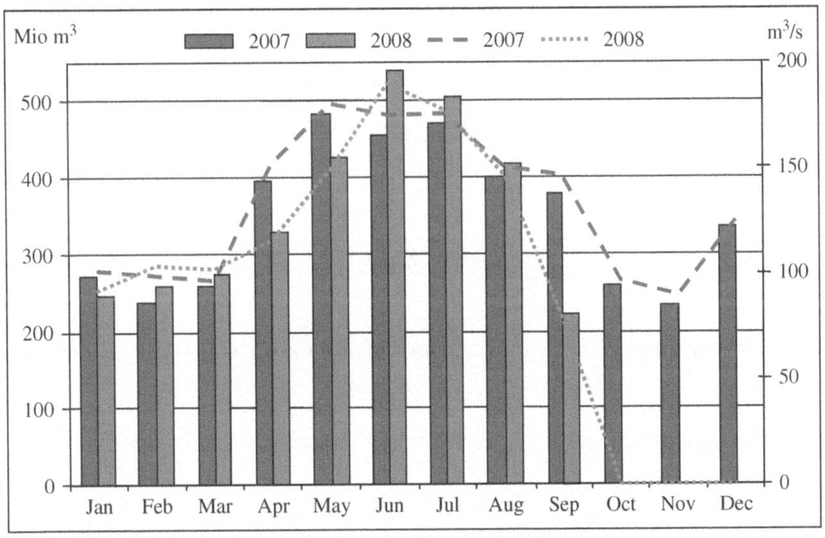

Figure 4.5. Water abstraction volume from Amu Darya to Karshi canal and its monthly distribution in 2007–2008.

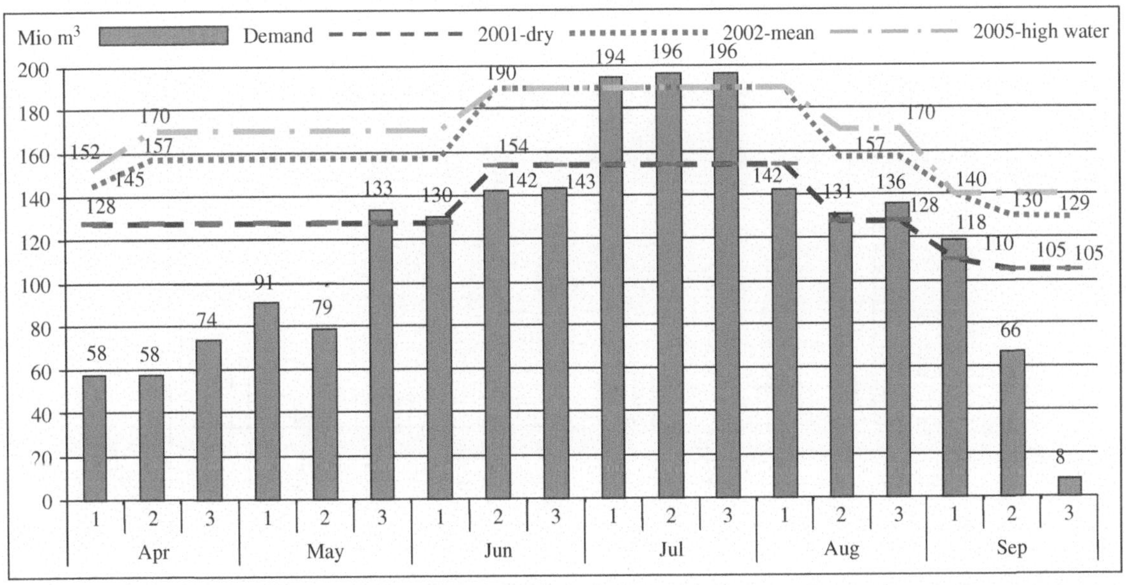

Figure 4.6. Comparison between demand and the limits in the Kashkadarya region for the areas irrigated from the Amu Darya.

4.1.3 Comparison of water supply and actual water use by irrigation projects

Water supply information accepted from the Water Ministry of Uzbekistan and BVO Amudarya were analysed for water demand and limits in dry, mean and high water periods. As an example two kinds of regions were considered:

- Regions taking water from the river only by pumping – Kashkadarya, Surkhandarya and Bukhara.
- Regions taking water by gravity – Khorezm amd Karakalpakstan.

The efficiency of irrigated agriculture in the regions, irrigated from different sources was analyzed:

- Surkhandarya: from Amu-Zang canal and Surkhandarya
- Kashkadarya: Karshi canal, Kashkadarya, Zarafshan
- Bukhara: from Amubukhara canal and Zarafshan

Detailed analysis of water use is provided for the Kashkadarya region in Figure 4.7. The Water Ministry is interested in minimising pumping from the Amu Darya covering as much of the water demand from other sources.

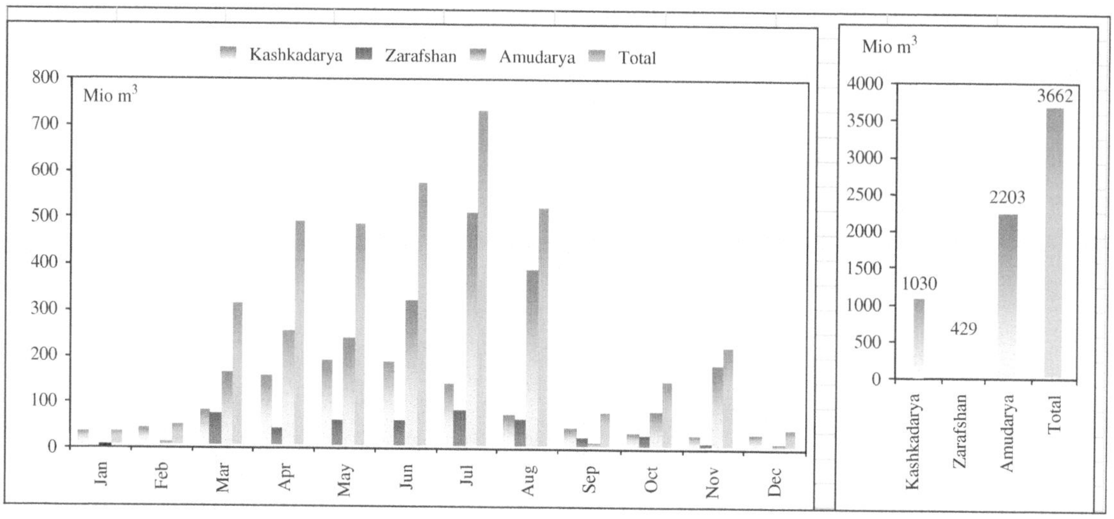

Figure 4.7. Average multiyear water consumption in Kashkadarya region (for period 1985–2007).

According to Figure 4.8 in 2008 over 50% of the water was used for cotton. In this region 46% of the irrigated plots (122,000 hectares) were under cotton, 40% were under wheat (98,000 hectares), 15% were under forage (37,000 hectares) and the remaining areas were used for vegetables and fruits.

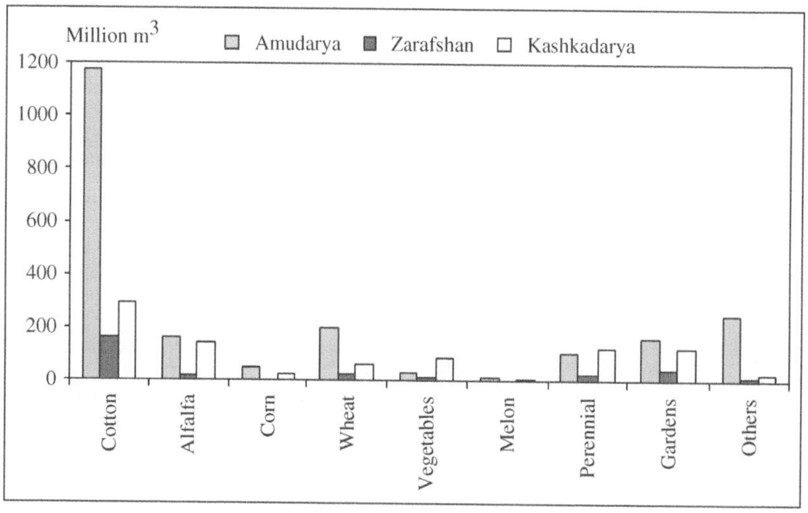

Figure 4.8. Water demand for irrigation of crops in Kashkadarya region in 2008.

In order to improve the management of irrigation systems, the project Jayhun, developed recommendations for urgent improvements of the water distribution plan. A water distribution schedule (plan) depends on the availability of water resources. An assessment of the actual demand of the water users required to produce planned crops was carried out and compared with the available water resources. The inflow and outflow was estimated in order to match demand to the available water. Two versions were considered:

- If the water resources are equal to or more than the estimated volume required. In this case the management unit water distribution plan is satisfactory.
- If the resources are less than the estimated demand, the water supply will be amended.

A method to urgently amend and redistribute the available water resources within the irrigation systems was developed for the special watering period taking into account the following:

1. The possibility to abstract additional water from other sources (in this case from the Zarafshan, wells, drainage from irrigation/or other waste water)
2. The available water resources were redistributed according to a system of priority between the irrigated areas. The first priority is to provide water for cotton. The priority was determined by the expert assessment of a special commission in order to reduce the water supply to lower priority areas. The assessment was based on an amount of damage expected due to water deficit.

The developed methods allow for increasing water use efficiency under systems of water shortage.

4.2 IRD-HYDRACCESS HYDRO METEOROLOGICAL DATABASE

The IRD team plans to use the Hydraccess software to compile and process the available data in a user friendly database. One of the reasons for choosing the Hydraccess software was the option to share it with all partners. Hydracess can be downloaded from the Internet and is free of charge. It is currently available in French, English and Spanish, and the IRD team have decided with the agreement of its author, Philippe Vauchel, to translate it into Russian, especially for the Jayhun project and its partners. This report provides a brief description of the software, and the existing Jayhun database.

Hydraccess is an extensive, homogenous and user friendly software, which allows the importation of various kinds of hydrological data into a Microsoft Access 2000 ® database[1], and to offer all the processing that Hydrologists require. It was created by a hydrologist for other hydrologists. Development began in the year 2000, and has regularly continued since. The author, Philippe Vauchel, is a hydrologist in the IRD (French Research Institute for Development), and the software remains the property of IRD.

Hydraccess exists in French, Spanish, English. A Russian version has been specially created for the Jayhun project. It can be downloaded freely at the following web address: http://www.mpl.ird.fr/hybam/outils/hydraccess.htm

4.2.1 Preliminary remarks

The network of hydrometeorological monitoring stations in the study area has been developed since the end of the 19th century, with a maximum number of stations in the period 1985–1986. Since then the degradation of the observation network has been continuous with most data series being disrupted in the 90s after the dissolution of the USSR. Though some stations re-started their activity in the 2000s, the Tadjik national hydrometeorological service has insufficient material and human resources to maintain the network. Automatic observation instruments are rarely used and the overall computerization of the system output is poor.

In addition it became rather difficult for the IRD team to obtain the existing long term data series on river discharges, temperature and precipitation. Some of the data received contained errors or were in unsuitable formats, and the geographical coordinates of the stations themselves were sometimes inexact. Even now, the database only contains the partial data which was received from project partners; this can be updated as more data becomes available.

For the project, some data from the re-analysis modelling approach available from NCEP NCAR (NOAA, United States) were downloaded in order to complete the database using external sources.

Finally meteorological data was retrieved from the "Global Historical Climate Network (GHCN) database: High-Mountain areas of Central Asia" publicly available on the internet (http://snobear.colorado.edu/Markw/Geodata/geodata.html), and this has been included in the Hydraccess Jayhun database.

[1] Hydraccess works also with the more recent versions of Microsoft Office (2003 and 2007).

4.2.2 Selected stations – Geographical coverage

The study area covers the region upstream of the Rogun dam for the Vakhsh and more particularly the Glacial regions of the Muksu, Kyzylsu, Obikhingou and part of the Piandzh (see Figure 4.9).

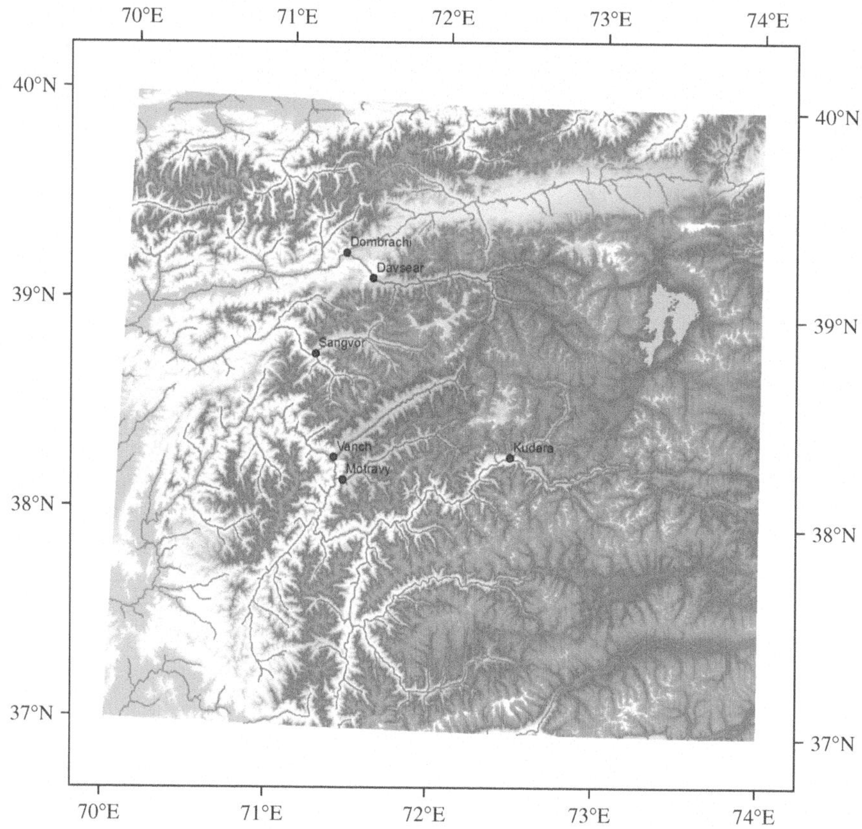

Figure 4.9 Location of the study area – in this map the topography and the water courses are established using the ASTER GDEM digital elevation model at 30m of ground resolution (product of METI and NASA, 2009). The outlets of the 6 benchmark basins used for the study are located on the map.

The stations were created in Hydracces through the 'Edit the Stations' menu. The station identification codes (ID codes) are those used by the Hydrometeorological Services. Comments, such as, for example, on doubtful or verified coordinates may be added.

The following tables provide the inventory of all the hydrological and meteorological stations entered in the database. Note that for the purposes of reanalysis, virtual stations were created for selected nodes within the grid (NMC1 to NMC4).

The coordinates of the measuring stations were checked and identified, with the assistance of Google Earth. Some coordinates of hydrological measuring stations had to be amended from the official values when it became apparent that they were incorrectly located in the mountains and not on a river. Subsequently, using the ArcGIS software, and on the basis of SRTM 90m digital elevation data (DEM)[2] the locations of selected hydrological stations were reviewed again (in order to be exactly located on the correct river stream) The following map (Figure 4.10) represents the location of selected stations on an ArcGIS map (some Meteo stations beyond the borders of the map are not represented, and only the main Hydrological stations required by the project team are indicated).

[2] Void-filled seamless SRTM data V1, 2004, International Centre for Tropical Agriculture (CIAT), available from the CGIAR-CSI SRTM 90m Database: http://srtm.csi.cgiar.org and http://www.ambiotek.com/topoview (Version3)

Table 4.1 List of Hydrological stations.

Station ID	2ndary ID	Name	Basin	River	Lat.	Long.	Alt. (m)	Basin area (km2)	Activity Start
17043	SH	Shidz (Шидз)	Piandzh (Пяндж)	Piandzh (Пяндж)	37	71	1954	57100	1962
17065	KU	Kudara Ust'e (mouth) (Кудара Устье)	Bartang (Бартанг)	Kudara (Кудара)	38,34	72,47	2600	4361	1947
17067	RU	Rushan (Рушан)	Piandzh (Пяндж)	Vomardara (Вомардара)	37,95	71,53	2049	87,1	1951
17068	MO	Kishl. Motravy (Киш. Мотравы)	Piandzh (Пяндж)	Yazgulem (Язгулем)	38,2	71,42	1803	1810	1932
17070	VA	Ust'e Vanch (Mouth) (Устье Ванч)	Piandzh (Пяндж)	Vanch (Ванч)	38,3	71,36	1610	2061	1964
17082	GA	Garm (Гарм)	Vakhsh (Вахш)	Vakhsh (Вахш)	39	70,34	1297	19936	1932
17084	KO	Darband - Komsomolabad (Дарбанд Комсомолобад)	Vakhsh (Вахш)	Vakhsh (Вахш)	38,86	69,98	1133	29308	1942
17100	DO	Dombrachi (Домбрачи)	Kysylsu (Кызылсу)	Kyzylsu (Кызылсу)	39,28	71,39	1833	8456	1960
17101	DA	Davsear (Давсеар)	Muksu (Муксу)	Muksu (Муксу)	39,18	71,56	2017	6500	1960
17102	YA	Yarmazor (Ярмазор)	Vakhsh (Вахш)	Pitaukul (Питаукул)	39,25	71,233		531	1960
17103	KA	Kapali (Капали)	Vakhsh (Вахш)	Churak (Шурак)	39,002	70,833		106	1970
17104	SM	Sangimaliki (Сангималики)	Vakhsh (Вахш)	Sarbog (Сарбог)	39,033	70,2	1257	1760	1950
17105	KM	Karamandy (Караманди)	Vakhsh (Вахш)	Kamarov (Камаров)	39,017	70,167		1700	1967
17106	SG	Sangicar (Сангикар)	Vakhsh (Вахш)	Sangicar (Сангикар)	39,033	70,002	1279	291	1945
17107	SV	Sangvor (Сангвор)	Obikhingou (Обихингоу)	Obikhingou (Обихингоу)	38,79	71,21	2183	1890	1950
17109	TA	Tavildara (Тавильдара)	Obikhingou (Обихингоу)	Obikhingou (Обихингоу)	38,712	70,525	1642	5323	1958
17109a	YO	Yozgand (Езганд)	Obikhingou (Обихингоу)	Obikhingou (Обихингоу)	38,717	70,4	1528	5651	2002
17111	OB	Obimazar kish. Sangvor (Обимазар)	Obikhingou (Обихингоу)	Obimizar (Обимазар)	38,77	71,22	2229	1002	1955
17113	KL	Kalaysang (Калайсанг)	Obikhingou (Обихингоу)	Saryob (Сариоб)	38,717	70,65		304	1961
17115	CK	Chakov (Шаков)	Obikhingou (Обихингоу)	Surkhsu (Сурхсу)	38,817	70,333		264	1961
17119	OBG	Obigarm (Обигарм)	Vakhsh (Вахш)	Obigarm (Обигарм)	38,714	69,712	1162		1962
17121	DG	Dagana (Дагана)	Vakhsh (Вахш)	Nurek (Нурек)	38,4	69,3			1971
17122	GO	Gofilabad (Гофилабад)	Vakhsh (Вахш)	Dagana (Дагана)	38,317	71,23			1963
17401	KH	Khait (Хаит)	Kysylsu (Кызылсу)	Yarkhitch (Ярхыч)	38,183	70,867		1150	1978
17459	TI	Tigrovaya Balka (Тигровая б.)	Vakhsh (Вахш)	Vakhsh (Вахш)	37,43	68,42		36200	1983

Table 4.2 List of Meteorological stations.

Station ID	Name	Имя	Lat.	Long.	Altitude (m)	Activity Start
BU	Bustanabad	Бустанабад	38,67	69,63	1983	1948
DO	Dombrachi	Домбрачи	39,15	71,567		1961
DU	Dushanbe	Душанбе	38,58	68,78	790	1926
FE	Fedchenko glacier	Федченко	38,83	72,22	4169	1935
FR	Fergana	Фергана	40,38	71,75	578	1881
GA	Garm	Гарм	39	69,95	1316	1935
HO	Horog	Хорог	37,5	71,5	2075	1899
KO	Komsomolabad	Комсомолабад	38,87	69,98	1259	1947
LA	Lyarun	Ляйрун	38,9	70,9	2008	1962
LY	Lyaksh	Ляхш	38,32	71,25	1998	1961
NMC1	NMC 72,5E - 37,5N		37,5	72,5		1948
NMC2	NMC - 70E 37,5N		37,5	70		1948
NMC3	NMC 70E - 40N		40	70		1948
NMC4	NMC 72,5E - 40N		40	72,5		1948
TA	Tavildara	Тавильдара	38,7	78,48	1616	1932

4.2.3 Sensors – Data sources

The sensors defined for each station correspond to each data series. A new sensor was created for the same station for each time series received. This was very helpful, in view of the laborious collection of different types of and often incomplete time series from the various partners.

Data from Hydromet sources

Though the original measured data belonged to Hydrometeorological services, elements were received in different formats and aggregations from the project partners. Generally for discharges, monthly data were available, though in some cases these were on a decadal or a daily time step basis. Hydraccess has a function to generate daily and/or monthly data series from this data, creating new sensors for the new time series. New sensors were created as well when data series were corrected in order to differentiate them from the original data.

Data from the GHCN database (High-Mountain areas of Central Asia)

This database provides information on meteorological observations (Air T° and precipitation) within the area 35°–50° N and 55°–85° E. As a general rule the sources of this data were the "Collections of Long Term Data or Meteorological Monthly Reference Books (Hydrometeorological services)". Most of these long term data series terminate between 1990 and 1994.

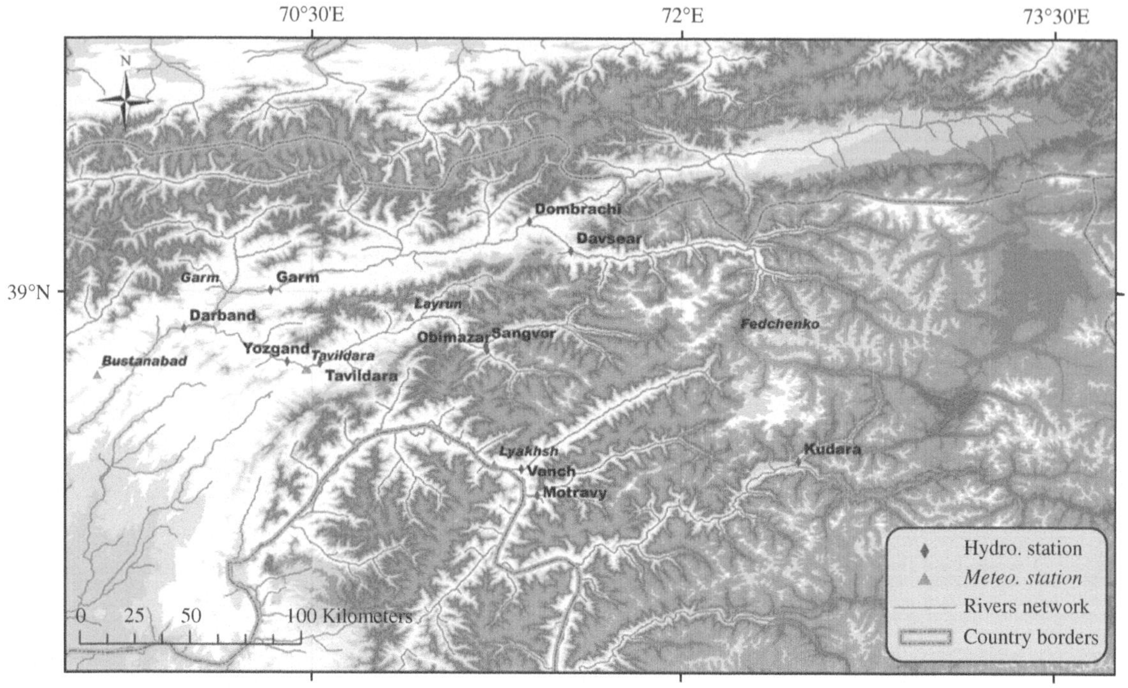

Figure 4.10 In this map the topography and the water courses are establish using the SRTM digital elevation model at 90m of ground resolution (enhanced by CGIAR, 2004).

Data from the reanalysis modelling approach from NCEP NCAR

The NCEP/NCAR Reanalysis Project is a joint project between the National Centres for Environmental Prediction (NCEP, formerly "NMC") and the National Centre for Atmospheric Research (NCAR). The project uses a state-of-the-art analysis/forecast system to perform data assimilation using data from 1948 to the present. A subset of this data has been processed to create monthly means. The spatial coverage is a global grid encompassing 2.5-degree latitude × 2.5-degree longitude. The Jayhun project area is located within the grid cell located between 70° E – 40° N and 72.5° E – 37.5° N.

The imported variables to the Jayhun database are: monthly air temperature (°C), relative humidity (%), specific humidity (g/kg), and precipitation (kg/m^2) .

The final version of the Hydraccess database can be downloaded free of charge from the following address: http://www.mpl.ird.fr/hybam/outils/hydraccess.htm

This includes a comprehensive tutorial in English and French (not translated into Russian).

Registered Jayhun project partners can download the Jayhun database for Hydraccess from the project website: http://www.inco-jayhun.net (-> Publications; -> Internal Documents; —> Data).

Others should submit a request to the project coordinator (Oliver Olsson, oo@warb.uni-hannover.de) in order to obtain access to the project database.

The data stored in the Hydracces database were provided by the Tadjik Hydrometeorological Agency, the Tadjik Academy of Sciences and the Russian Academy of Sciences. Data were also acquired from the reanalysis models from NCEP NCAR (NOAA, United States) and meteorological data from the "Global Historical Climate Network (GHCN) database: High-Mountain areas of Central Asia".

Any use of the Hydraccess database and software should make reference to IRD.

4.3 RIVER INFLOW PARAMETERS AT THE THC BY THE HYDROMETRICAL STATION

An automatic sensor was installed as an inflow reference station for the THC reservoirs (Figure 4.11) recording short-term turbidity dynamics in the lower Amu Darya River.

New data on salinity, turbidity, water level, and temperatures is available for the period 2007 and 2008 (January – December) establishing new inflow reference values for the reservoirs of the Tuyamuyun Hydro-Complex. An additional unit for the synchronization and transmission of data through a cellular network (mobile phones) was also installed. The data are continuously transferred and are now available in a newly created database with a user interface. This includes an analysis tool and a user input-output interface (Figure 4.12).

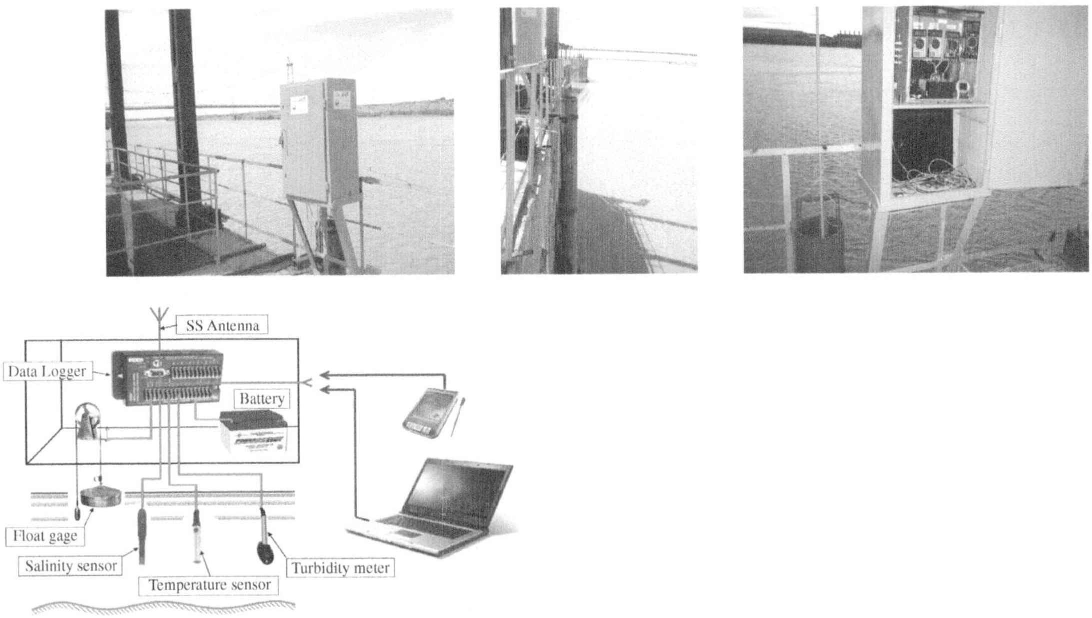

Figure 4.11 The new hydrometrical station installed by the Jayhun project.

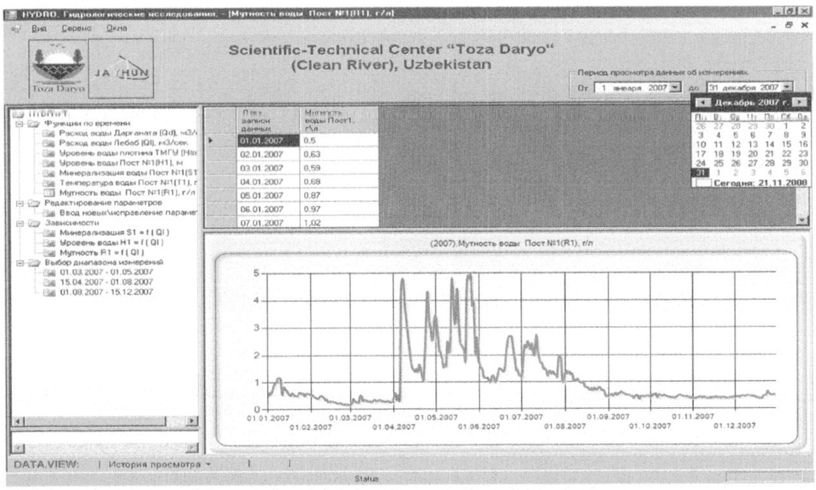

Figure 4.12 User's input-output interface for the Hydrometrical station database.

Chapter 5

WEAP – Modelling of Amu Darya water resources

David Rycroft

5.1 DEVELOPMENT OF THE AMU DARYA WEAP MODEL

The main aim of this work package was to identify a modern water resource management programme that allows options for the operation and management of the Amu Darya river system to be explored, enabling planners to investigate the impact of changes in management practices. This work package developed a water resource management model for the Amu Darya basin using the commercially available programme WEAP (but free to Developing countries). The model developed in stages. The first stage involved the development of a model based on 25 years of recorded flow data; this is named AMU-WEAP. The second extended the period of application to 100 years using flow data generated from the 25 year record by stochastic analysis. The key findings are all based on the output from the 100 year version of the model, named AMU-WEAP_100.

The software package WEAP (Water Evaluation and Planning System) was developed by the Boston Centre of the Stockholm Environment Institute. WEAP integrates the demand and supply side of water use, river discharge, reservoirs and water transfers. The programme is comprehensive, straightforward and most importantly is intuitive and easy-to-use. It attempts to assist rather than to substitute for the skilled investigator and has been successfully applied throughout the world to investigate alternative water management strategies.

5.1.1 The principles of WEAP

WEAP operates on the principle of water balance accounting. Model development starts with the creation of a schematic of the basin, based on its various sources of supply and demands (Figure 5.1). Sources include rivers and groundwater. Demands include water for irrigation and for human and/or industrial use.

The sources are connected to the centres of demand by *transmission links* (in reality irrigation canals or pipelines). *Return flow links* (in reality drains) convey excess water back, either to the rivers, to reservoirs or to other points of disposal. Rivers within the schematic start at the most upstream flow gauge on each tributary and end, in the case of the these models, at the Aral Sea.

Demands for irrigation are considered by introducing *catchments* into the schematic. Demands for urban and industrial water use are considered by introducing *demand sites* into the schematic. Figure 5.1 is a schematic of the upper Amu Darya basin with the tributaries: Pyandj, Vaksh, Kafirnigan in Tajikistan; the Kunduz and Kokcha in Afghanistan and the Surxondaryo in Uzbekistan.

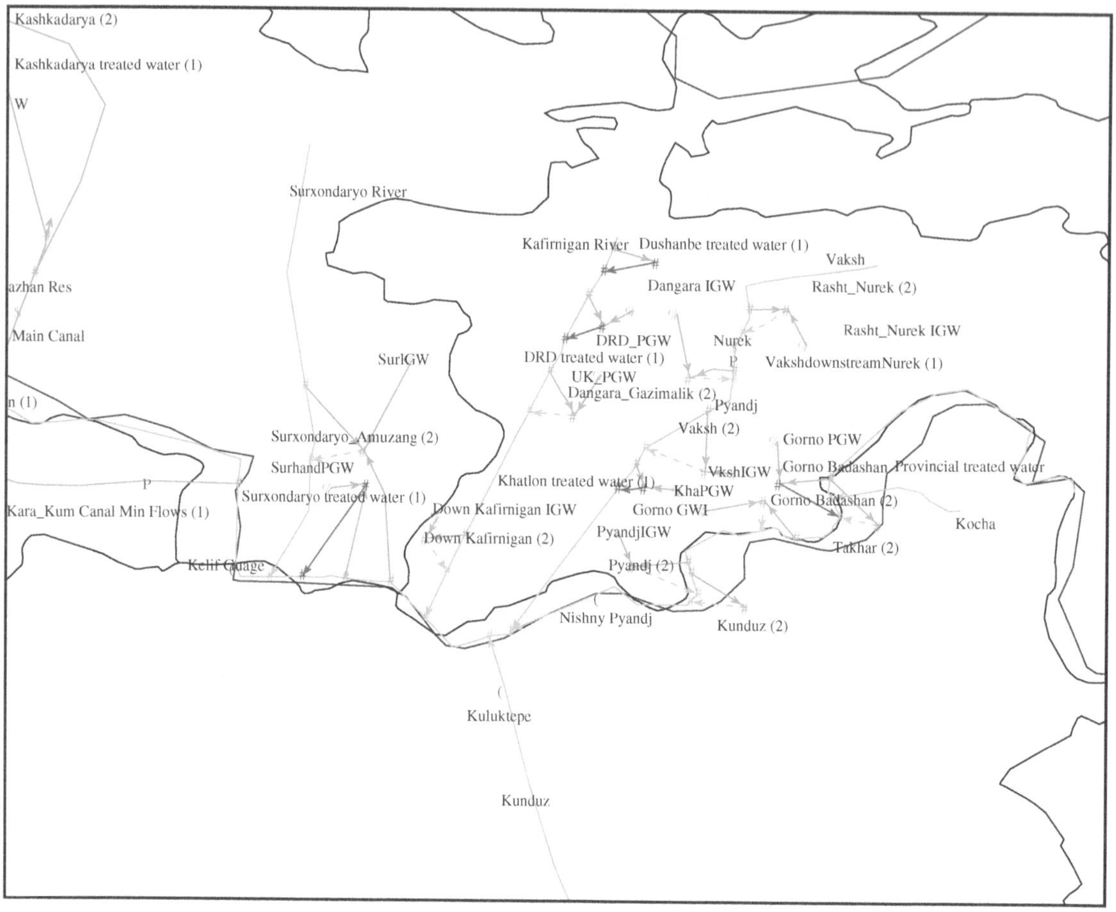

Figure 5.1. A Schematic depicting the Upper Amudarya basin.

Data input

Data input starts with the creation of a baseline scenario for one particular year; this is known as the current accounts year and it contains all of the fundamental structural data for the model (schematic, river discharges, reservoirs, irrigated areas, populations etc.). A separate and quite independent reference years scenario is then created containing the data for all other years: e.g. 1981–2005. Other scenarios can then be created to consider e.g. the effects of climate change or reductions in river flows.

River discharges are key elements of the data, normally input in the form of a time series of the known monthly discharges at the furthest upstream gauge on each tributary. In WEAP this is known as the "*headflow*". Selected patterns of monthly or annual flow can also be imposed, and flow gauges included to allow comparison between the modelled and the recorded discharges.

Groundwater use can also be simulated, necessitating data on the total quantity of groundwater storage available, the rates of recharge to and the rates of withdrawal from groundwater.

Data needed to simulate surface reservoir operation includes the storage-elevation relationship. Zones considered within reservoirs include: from top to bottom, the flood-control zone (e.g. temporary storage above a spillway), the conservation zone, the buffer zone and the inactive zone. The conservation and buffer zones together constitute a reservoir's active storage. WEAP allows water to be released without restriction from the active zone in order to meet any demands arising downstream. Once the water level enters the buffer zone, releases may be limited, e.g., by only allowing 5% of the remaining buffer zone storage to be released each month. No releases are possible once the water level enters the inactive zone. Hydropower operation can also be simulated, necessitating additional information on the level of the turbine outlets and the system of power generation.

Demands: irrigation

The menu option *catchments* has been selected for irrigated areas since it includes routines to determine crop water requirements. The names of the irrigated catchments within Tajikistan are: Rasht-Nurek, Dangara-Gazimalik, Vaksh, Up Kafirnigan, Down Kafirnigan, Pyandj and Gorno-Badashan. The irrigated catchments within

Uzbekistan are: Surxondaryo-Amuzhang, Kashkadarya, Navoi, Bukhara, Upradik-Karakalpakstan, Lower BVO Karakalpakstan and Khorezm. The irrigated catchments within Turkmenistan are: Mary, Ahal, Lebab and Dashkauz. A single catchment has been included in Afghanistan (Afghan-Pyandj) to allow existing and future water use to be considered. The total irrigated area (1980–2005) was approximately 3.4 million ha.

The "FAO CROPWAT" routine has been used within the WEAP programme to determine crop water requirements (FAO, 1998). This takes climatic data (monthly precipitation and monthly reference crop (*grass*) potential evapotranspiration, ET_o) and combines it with information about the areas of irrigated land and the range of crops being cultivated to determine the *potential evapotranspiration* by the crops (ET_p). The individual elements of potential evapotranspiration are then summated and increased to allow for irrigation efficiencies (water lost in the delivery and application of water as well as return flows) in order to establish the overall *demand* for water.

All catchments are connected to the sources (i.e. a river or groundwater) either directly by a transmission link or indirectly by a diversion connecting to a transmission link leading to the irrigated area. All catchments include a return flow link to convey water back, either to the river, or to some other point of disposal.

Data for irrigated areas was provided by our partners, and was supplemented where necessary by information gathered from secondary sources (EU reports, satellite imagery and databases, e.g. FAO: AQUASTAT, 2008). Essential data includes: the irrigated area (in WEAP termed *annual activity*) the range and the distribution of crops; the cropping calendar, the crop growth factors as well as the irrigation efficiencies (in WEAP termed the *irrigation fraction*).

Required climatic data includes: the mean monthly precipitation as well as the mean monthly reference crop (grass) potential evapotranspiration (ET_o).

Demands: treated water

The menu option demand sites has been selected for treated water supplies. A single demand site has then been used to represent the water use in an entire province within the country concerned (confined to the Amu Darya basin). Key data includes the population (in WEAP termed *the annual activity*), the division between the urban and rural populations, the industrial share of water and the annual consumption per water consuming category. Information about the losses of water in both transmission and delivery is also included.

All provincial treated water demand sites are connected to the sources (i.e. rivers or groundwater) by transmission links. A single return flow link then conveys water back, either to the river, or to some other point of disposal. Losses in supply to provincial treated water demand sites have been applied through the return flow links.

The population (in 1991) was adopted for all provincial treated water demand sites. These include four provinces in Tajikistan: Khatlon, the Direct Rule District (DRD), Dushanbe and Gorno-Badashan. In Uzbekistan they include the provinces: Surxhondaryo, Kashkadarya, Navoi, Bukhara, Karakalpakstan and Khorezm. In Turkmenistan they include:- Mary, Ahal, Lebab and Dashkauz. No allowances have been included for population growth.

The conveyance infrastructure: transmission links and diversions

Water is conveyed to demand sites by means of transmission links. These can either connect the source and the demand site directly, or connect indirectly to a diversion which takes it's water directly from the source (diversions provide greater flexibility for managing the internal flows). Losses in supply and distribution to irrigated areas have been applied as a notional loss of water from the transmission link. In WEAP diversions are considered to be functionally identical to rivers, providing much greater scope for managing discharges through them than is possible using transmission links.

Demand sites can be assigned different *demand priorities* for the supply of water. All provincial treated water demand sites have been assigned the highest demand priority of 1, whereas all irrigated catchments have been assigned lower demand priorities of either 2 or 3. This ensures that, in cases of water shortage, treated water demands are met before any water is allocated to irrigation.

Water can be supplied by surface water sources alone, by groundwater sources alone or by a combination of both by assigning different *supply preferences* to the transmission links leading to them. Assigning a supply preference of 1 to a transmission link from a groundwater source will ensure that water is abstracted first from the groundwater source. Water will only be abstracted from a lower preference source (e.g. a river) once the stated maximum withdrawals have been taken from the higher preference source.

WEAP endeavours to supply water to all demand sites, subject to there being sufficient water available at the source (i.e. the inlets to the conveyance infrastructure). However, if having satisfied priority 1 sites (i.e. treated water supply) shortages remain, WEAP will apply the shortages equally across the next lower demand priority sites (i.e. irrigated areas).

The supply to a particular demand site may be constrained by imposing a limit on the maximum capacity of the diversion or transmission link leading to the site. The controls available to manage flows within diversions include the ability to impose both maximum and minimum values; applying identical maximum and minimum values allows selected patterns of flow to be imposed.

Demand sites generate their own demands for water, either for treated water or for irrigation. Satisfying these depends firstly upon there being sufficient water available at the sources. However, if capacity is constrained, e.g. by a limit on the maximum capacity of the transmission link or diversion, then notwithstanding the availability of water, it may be not be possible to satisfy demand. In such circumstances *shortfalls* will arise.

Shortfalls (for irrigation) are reported in two different formats: *ET shortfall* (ETs) and *Unmet demand*. ET shortfall is the water needed to bridge the gap between the potential evapotranspiration for the crops (the water the crops require for optimum growth) and the actual evapotranspiration (influenced by the supply constraints). The unmet demand, however, is the quantity of water, which needs to be abstracted from the sources to satisfy the ET shortfall, and this generally exceeds the ET shortfall because it includes additional water needed to satisfy irrigation efficiencies and return flows.

Output

WEAP provides a highly detailed presentation of results. These can be displayed for all demand sites, or for any selection. The results for all years, selected years, as well as all months or selected months can be displayed, either in graphical or tabular format. Statistics and routines allow the results to be exported to EXCEL or WORD format. In WEAP, results are categorised under the following headings:

- *Demands include*: Water Demand; Supply Requirement; Supply Delivered; Unmet Demand; Coverage; Reliability; Demand Site Inflows and Outflows; Instream Flow Requirements.
- *Supply and Resources include*: Inflows to Area; Outflows from Area; *River* [further sub-divided into: Streamflows: Streamflows Relative to Hydropost %, Stage, Velocity, Reach Length];
- *Groundwater* [further sub-divided into: Storage, Inflows and Outflows, Overflows, Outflows to River];
- *Reservoirs* [further sub-divided into: Storage Volume, Storage Elevation, Inflows and Outflows, Hydropower];
- *Transmission Link* [further sub-divided into: Discharge, Inflows and Outflows];
- *Return Link* (further sub-divided into: Flow, Inflows and Outflows);
- *Catchments include*: Land Class Inflows and Outflows, Runoff and Precipitation, Observed Precipitation, Infiltration/Runoff Discharge, ET Potential, ET Actual (including irrigation), ET Shortfall;
- *Input Data includes*: Key Assumptions, Demand Sites and Catchments, Supply and Resources and Water Quality.

5.1.2 Model evolution

Two different versions of the model were developed. The first, AMU-WEAP, was based on 25 years of recorded flow data for the principal tributaries of the Amu Darya. This was used to adjust parameters, e.g. losses in transmission, in order to match the recorded and the modelled river flows.

The second, AMU-WEAP_100, was developed to generate a longer and statistically more sensible time series (100 years) and to help overcome concerns about the reliability of the recorded flow data in some of the 25 years. The fundamental structure of both models is identical, although some changes were introduced into AMU-WEAP_100 in order to improve model simulation. All results are based on the second stage model AMU-WEAP_100.

Calibration

Calibration sought to achieve agreement between the recorded discharges at the various gauging stations along the Amu Darya and the discharges as determined by the model. The available data included monthly discharges for the principle tributaries of the Amu Darya for: 1980–2005 but data on the factual abstractions was only available for the 15 years:-1991–2005. Calibration was therefore restricted to the 15 years of combined data.

Achieving agreement proved challenging due to inconsistencies between the various data sets, including the areas of land supplied from particular canals and the abstractions from those canals. Inspection of the raw data revealed, for example, that the differences between the flows at the Kelif gauge (the furthest upstream on the Amu Darya) and those at Kizildgar (the closest to the Aral Sea), which in theory should equate to the abstractions between these two gauges did not. The flows calculated for Kizildgar by subtracting the recorded abstractions from the flows at Kelif were 12 km^3 higher than the recorded values.

The factual abstractions record was inspected in order to identify the highest monthly discharge in the 15 year record, either in an actual named canal or in a notional "aggregated" canal supplying the areas: Khorezm, Upradik_Karakalpakstan, Lower BVO Karakalpakstan, Lebab and Dashkauz. These values were then used to impose maximum capacities on each of the diversions conveying water to the relevant irrigated areas. The Karakum Canal was dealt with in a slightly different manner since it became apparent that the maximum capacity of 600 m^3/s only rarely occurred; and further that adopting such a high maximum led to diversions that were much greater than the recorded values. Accordingly in this case values of the mean monthly diversions were imposed, based on the 15 years of factual records, by using identical maximum and minimum values.

The most significant parameters used to adjust the flows along the Amu Darya were the losses in delivery to the demand sites and the losses due to seepage from the bed of the Amu Darya.[1] The losses from the irrigated areas had the far greatest influence. These were all based on the officially reported irrigation efficiencies (SANIIRI for Uzbekistan and FAO AQUASTAT for Tajikistan and Turkmenistan). The initial working premise assumed that all water diverted in excess of crop water requirements was lost within the system. The value used for the losses were then adjusted to ensure that 20% (Tajikistan) or 10% (Uzbekistan) of the diverted water returned to the river. This reflected the fact that irrigated areas in Tajikistan are located in valley type situations favouring return flows. Irrigated areas in Uzbekistan are typically located far from the river with much less favourable opportunities for return flows. No allowance for return flows have been included for the two large irrigated areas in Turkmenistan, Mary and Ahal, supplied from the Karakum canal due to absence of any obvious visual return flow links to the Amu Darya.[2]

The output from this model yielded satisfactory agreement between the predicted and the recorded annual discharges at the various gauging stations, particularly at Kelif and Kizildgar, both for the 1991–2005 record relied on for calibration and the extended 1980–2005 record incorporated into the model AMU-WEAP. The predicted mean annual discharge at Kizildgar for the 25 years was 8.30 km^3 compared to the recorded value of 8.36 km^3. The discharges at Kelif were (predicted = 61.88 km^3, gauged = 61.22 km^3), and at Tuyamuyun (predicted = 29.60 km^3, gauged = 29.63 km^3).

Differences remained between discharges at other gauges located below the Tuyamuyun complex (e.g. at Kipchak and Samanbay) but this the result of consolidating the abstractions from many outlet points into single aggregated abstraction points; and was not considered to be an issue affecting the overall balances within the basin.

5.2 SCENARIO MODELLING AND RESULTS

5.2.1 The model: AMU-WEAP_100

In order to evaluate the impact of climate change it was considered desirable to develop a longer time series than the 25 years used in the first stage model AMU-WEAP. The model AMU-WEAP_100 is an extension of the model AMU-WEAP but applies over a period of 100 years.

Substantive differences between the models AMU-WEAP_100 and AMU-WEAP

The fundamental structure of both models:- AMU-WEAP and AMU-WEAP_100 is identical.

Differences include:

(a) AMU-WEAP_100 relies on 100 years of flow data for the principal tributaries of the Amu Darya, generated from the 25 years of historical data (1980–2005) but using standard stochastic procedures. It applies over a period of 100 years, purely for identification purposes labelled as: 2001–2100.

(b) AMU-WEAP adopted single values for the maximum capacities of the diversions (with the exception of the Karakum Canal where maximum and minimum monthly averages were imposed); AMU-WEAP_100 extended this process by applying maximum canal capacities of diversions on a monthly basis, with values based on the mean monthly abstractions from the 15 years of recorded data. This had the effect of imposing an upper limit on their conveyance capacities.

(c) AMU-WEAP_100 does not allow for the growth of populations or of irrigated areas since such time related changes, when considered over the longer time periods (100 years) lead to unrealistic changes. The areas assigned to the irrigated catchments in AMU-WEAP_100 are those generally applicable in 2005 (SANIIRRI). Some allowance for growth of irrigated areas was included in the first model AMU-WEAP where information was available e.g. the Bukhara irrigated area increased in size from 235,723 ha in 1980 to 273,750 ha by

[1] Ivanov, E.G., and G.Kh. Ismailov (1995). Assessment of Natural and Anthropogenic Losses of the Amu Darya Runoff in Upstream, Middle Course, and Downstream Sections of the River. *Water Resources*, Vol. 22, No. 1, pp 1–8.

[2] Due to the way in which the WEAP programme has been constructed, losses have necessarily had to be imposed on the transmission links leading to each of the irrigated areas. For example, irrigation efficiency in Tajikistan is reported to be 72% (FAO AQUASTAT). Thus for every 100 mm of water consumed by the crops, 138.8 mm needs to be abstracted (100/0.72). The excess (38.8 mm), corresponding to 28% of the diversion, was initially applied as a notional loss from the transmission link. Subsequently an included "*Irrigation Fraction*" of either 80% (Tajikistan) or 90% (Uzbekistan) ensured that 20% or 10% respectively of the abstracted water returned to the river. The loss of 38.8 mm, including an allowance for 10% return flows in Tajikistan, culminated in an adjusted loss factor of 25% (rather than 28%) being used in the model AMU-WEAP_100. The actual loss within the irrigated area remains as: 38.8 mm, and the return flows as:15.4 mm (10% of diversion).

2005. This slightly different approach accounts for the slightly larger overall area being allocated to irrigation in the model AMU-WEAP_100, in turn accounting for the slightly higher estimates of demand of water for irrigation.

(d) The losses from irrigated areas were all amended to allow for the return flows (10% or 20%).

(e) The initial output revealed that the model (AMU-WEAP_100) was not mirroring the abstractions during the non-vegetated season (October to March). This is because WEAP is demand driven, primarily by irrigation, which is of course mainly a summer activity. In order to resolve this issue, the "relatively low" mean monthly abstractions during the non-vegetated season were imposed on the diversions leading to the irrigated areas in both Uzbekistan and Turkmenistan (but not Tajikistan due to the absence of relevant information). This strategy ensured that the abstractions were accounted for despite demands being extremely low. During summer, when demand becomes very high, the diversions were left free to respond to the needs of irrigation and water supply, subject only to their capacity being constrained by maximum capacity limits (April to September) and the water available at the sources.

5.2.2 Climate change

The model AMU-WEAP_100 includes eight scenarios. The *current account* year is 2001. The *reference scenario* applies to the *existing* climate (in 1961–1990, shortened in the text to 1985). The scenario "Climate 2085" applies to the predicted median climate in 2070–2099 (shortened in the text to 2085). The timescale adopted, 2001–2100 is purely notional and is only used for identification purposes.

The median rise in temperatures due to climate change (1985–2085) were derived *from the most recent Inter-Governmental Panel on Climate Change (IPCC) 4th Assessment Report*, for the scenario judged to most closely correspond to recent patterns of climate change in Central Asia (the A2 storyline). This assumes a very heterogeneous world, with high population growth and slower economic growth than other climate change storylines.

The analysis considered firstly the existing climate (in 1985) using downscaled climatic data derived from the *International Water Management Institute's 2007 database* (IWMI, 2007) but based on data generated by the *Climate Research Unit of the University of East Anglia* (CRU data) specific to each of the irrigated areas. Secondly, downscaled climate data for 2070–2099 (2085), also derived from the CRU but based on the most recent 4th IPCC assessment report. The initial consideration of climate change was confined to the median estimate of climate change and the (SRA2) climate change scenario.

Temperature variations across all global circulation models within the IPCC 4th assessment

The variations of temperature predictions across the entire range of global circulation models considered by the 4th IPCC assessment were reviewed in order to identify the spread of temperature variation between models for the SRA2 climate change scenario. This indicated that a monthly variation of plus and minus 2.5°C, relative to the median values for 2085, accounted for all significant variation.

The impact of this range of temperature variation on the reference crop potential evapotranspiration (ET_o) was then determined independently using the FAO CROPWAT routine (Clarke *et al.*, 2000). This indicated that the ET_o values by 2085 would range between plus 5.5% and minus 5.5% relative to those values based on the median climate change predictions. The revised estimates of ET_o were then introduced into two new scenarios based on, respectively, an increase of 5.5% and a decrease of 5.5% in the median values of ET_o for 2085.

Three scenarios considering reduced river flows were then introduced, all adopting the median (SRA2) climate predictions for 2085 but then assuming that all the tributary headflows decreased by respectively: 10% , 20% and 30% compared to existing values (the reference scenario) due to adverse changes within their catchments (e.g., glacial shrinkage/altered patterns of runoff).

5.3 FINDINGS

5.3.1 River discharges: the existing climate (in 1985)

The annual flows in the six principal tributaries of the Amudarya: the Pyandj, Vaksh, Kafirnigan, Kunduz, Kokcha and Surxhondaryo, are depicted in Figure 5.2, and tabulated in Jayhun report D.5.2. The mean annual combined flow was 66.2 km^3, the minimum (notionally in the year 2015) 49.1 km^3, and the maximum (notionally in the year 2099) 89.0 km^3. These it should be noted were derived by stochastic analysis from the historical data: 1980–2005.

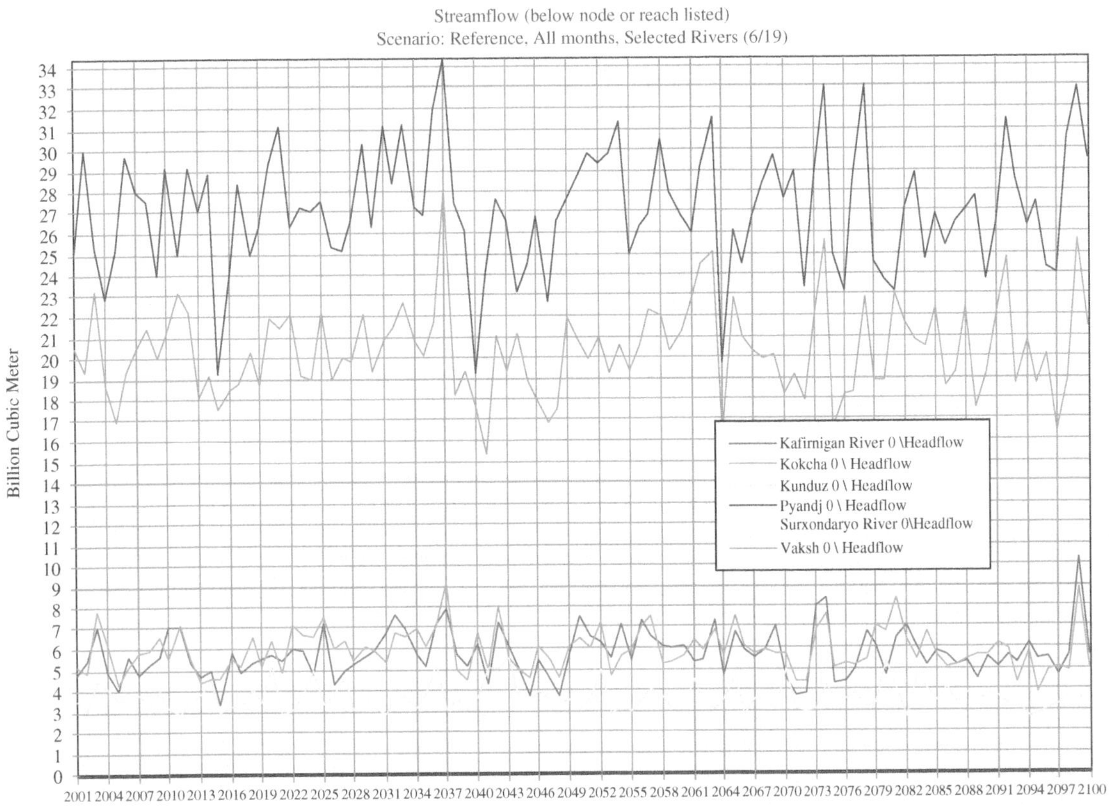

Figure 5.2. The annual headflows for the tributaries: Pyandj, Vaksh, Kafirnigan, Kunduz, Kokcha and Surxhondaryo, billion m^3 (km^3).

5.3.2 Evapotranspiration

The mean annual *potential evapotranspiration* (ET_p) for all irrigated areas for the existing climate (1985) was 34.5 km^3. The mean annual *actual evapotranspiration* (ET_a) for all irrigated areas, was 25.9 km^3. This is less than the potential due to the supply constraints; these include insufficient capacity in the conveyance systems (e.g. the irrigation canals) as well as shortages of water at the sources in some years (e.g. the rivers).

Values of potential evapotranspiration estimated for the climate in 2085 range from a low of 37.59 km^3 to a high of 41.21 km^3. The median is 39.26 km^3, which is an increase of 14% on existing values (1985).

The mean *annual shortfall* of evapotranspiration (ET_s) was 8.56 km^3. Shortfalls did not vary significantly between high and low flow years due to the overarching influence of the conveyance capacity limits, and not (as might be expected) by the shortages of water at the sources. This indicates that, even under the existing conditions, there are shortages and an implied inability to irrigate some 25% (8.6/34.5%) of the existing irrigated area.[3]

Values of the estimated shortfalls for the climate expected in 2085, assuming the existing river flows remain the same, range from a low of 10.80 km^3 to a high of 13.68 km^3, and a median of 12.13 km^3. Expressed as percentages they correspond to: 29%, 33% and 31% of the potential evapotranspiration.

The shortfall, based on the median climate predicted for 2085, if tributary headflows were to decrease by 20%, rises to 16.12 km^3, corresponding to some 41% of the potential.

Figure 5.3 depicts linear relationships between the combined annual tributary headflows and the ETshortfalls (ET_s) for the existing climate (1985) as well as the median predicted climate for 2085. The shortfalls are, as would be expected, less during years of high flow but they do not decrease to zero (as might be expected) due to the previously mentioned constraints on conveyance capacity. The minimum value of ~7 km^3 (1985 conditions) reflects the inability of the existing conveyance infrastructure to deliver sufficient water to meet demands, despite the availability of water at the sources, e.g. the entrances to the conveyance structures.

[3] The ratio ETshortfall/ETpotential has been used to identify the likely areas of land, which it will not be possible to irrigate. This ratio should be regarded as being indicative only since other factors, such as diversions during the non-vegetative season will contribute to reducing the shortfalls.

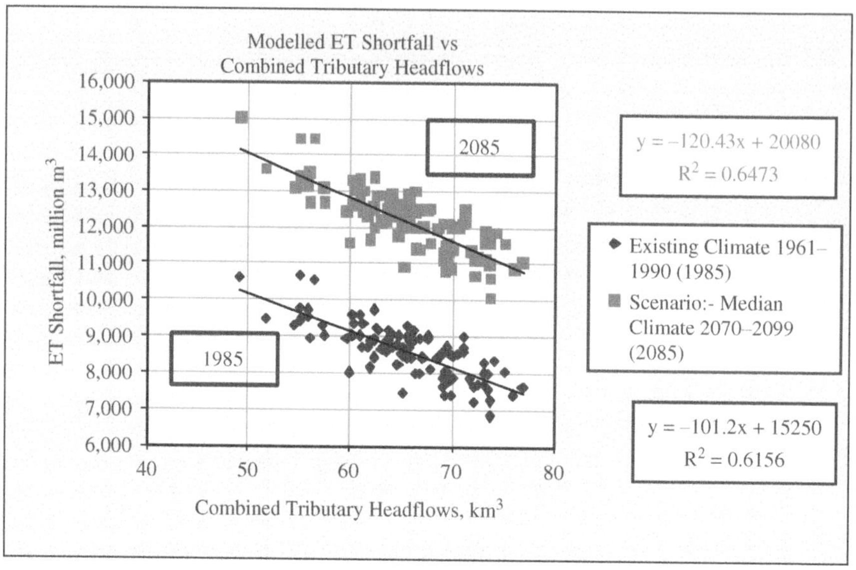

Figure 5.3 The combined annual tributary headflows versus the annual shortfall of evapotranspiration for the existing climate (1985) and the median predicted climate for 2085.

The model AMU-WEAP_100 was then modified in order to remove all constraints on conveyance capacity. The findings, illustrated in Figure 5.4, also depict linear relationships between the headflows and the shortfalls, both for the existing climate (1985) and the median predicted climate for 2085. In this case the shortfalls arise solely because of a shortage of water at the sources (e.g. at the entrances to the conveyance infrastructure).

Figure 5.4 reveals that the shortfalls do, as anticipated, now decrease to zero as the combined tributary headflows rise. More significantly, there are no shortfalls when the combined headflows exceed ~72 km^3. Beyond this combined river flow, the actual and the potential crop evapotranspiration are identical (at 34.5 km^3) which implies that the overall basin water use efficiency is ~52%.[4]

5.3.3 Abstractions

Irrigation

Abstractions to irrigated areas are derived from rivers as well as from groundwater sources. Each demand site receives it's water via one or two transmission links. The flow along each transmission link to irrigated areas includes water, which is ultimately consumed within the system (supply and distribution losses, consumption by the crops as well as return flows). Note: losses of water to provincial treated water demand sites are assigned to their return flow links.

The total quantities of water *abstracted for irrigation* (including water needed to satisfy irrigation efficiencies, consumption and return flows) for the existing climate, assuming capacity constraints apply, is detailed in Table 5.8. The mean abstraction was 44.7 km^3. Values did not vary significantly between low flow and high flow years because of the effects of the constraints on conveyance capacity. The loss of water was 17.9 km^3, the volume delivered to the irrigated demand sites 26.8 km^3, the actual consumption by the crops (ET_a) 25.9 km^3, the contribution of precipitation to consumption 2.48 km^3 and the return flows 3.31 km^3.

Estimates of the quantities to be abstracted for irrigation by 2085 range from a low of 46.07 km^3 to a high of 47.6 km^3, and a median of 46.88 km^3. The estimated abstraction by 2085 if the flow in the tributaries decreases by 20% is 39.2 km^3, reflecting the increasing impact of the shortages of water in the rivers.

[4] In the (notional) year 2050 the combined annual headflow was 72 km^3. The residual flow entering the Aral Sea was 8 km^3; the potential evapotranspiration (and consumption) was 34.5 km^3, the shortfall was zero. The consumption of treated water was 3.1 km^3. The total consumption was therefore 37.6 km^3. The magnitude of the unaccounted water was 26.4 km^3. The ratio between the consumption (37.6 km^3) and the net resource 72 km^3 is 52% which is an indirect measure of the basin's overall water use efficiency.

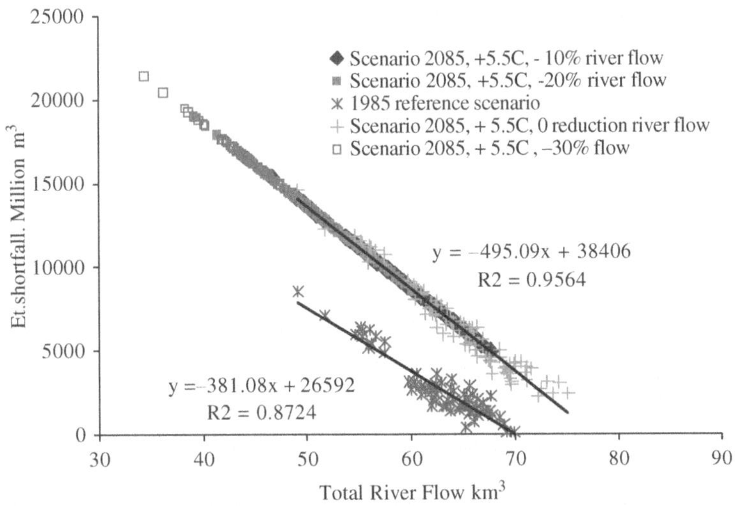

Figure 5.4. The effect of global warming with a median temperature rise of 5.5°C combined with reduced river flows on the ability to meet the basin's annual mean crop water requirements (ET_p) in 2085 (Based on the model 'AMU-WEAP_100' assuming there are no infrastructure capacity constraints). (Each point represents an annual flow from within the 100 years data set). Existing water use efficiencies are assumed.

All demands

The quantities of water abstracted to supply all centres of demand (irrigated areas plus treated water but excluding two included drainage lakes) were determined (JAYHUN report D.5.2). The mean existing annual abstraction (1985 values) was ~49.3 km^3 but by 2085 this is expected to rise from a low range value of 50.7 km^3 to a high of 52.2 km^3 and a median of 51.5 km^3.

Demands for treated water: the existing climate (1985)

The mean annual quantities of water abstracted for treated water purposes for the existing climate are detailed in JAYHUN report D.5.2. The mean abstraction was 4.6 km^3, corresponding to some 10% of all abstractions. Estimates of demand for treated water do not vary with time since they are unaffected by climate change and there are no allowances included for population growth.

Water allocation

Figure 5.5 illustrates how water has been allocated within the WEAP model. All values are in cubic kilometres and are means derived from the 100 years of data. The combined annual tributary headflow was 66.2 km^3. *This represents the entire surface water resource available*. An additional 1.57 km^3 was derived from groundwater (although this was most likely to have been recharged from surface water resources within the basin).

The actual consumption of water by the crops, by humans and industry was 29.36 km^3. Effective precipitation accounted for some 2.47 km^3 of consumption. Losses in transmission, distribution and application within the irrigated areas amounted to 17.91 km^3. Other losses within the basin include losses in the distribution of treated water (0.6 km^3), seepage from the bed of the Amudarya (3.21 km^3) and direct evaporation from the two included drainage lakes (0.74 km^3).

5.4 IMPROVING SIMULATION

As explained, the simulation of water use during the non-vegetated period was unsatisfactory. This reflected the fact that there were no procedures available within the WEAP programme to account for water diverted in one period e.g. February (as is currently used to pre-wet soils) but which is consumed at a later date (e.g., by crops from March onwards). The issue was resolved by imposing the (*relatively low*) abstractions throughout the non-vegetated season (October to March) on the diversions used to convey water to the irrigated areas.

Demand is low during the non-vegetated season, driven mainly by water needed by humans and industry as well as water used to pre-wet and leach the soils. One consequence of the decision to impose abstractions is the emergence in results of "*residual flow*" from the diversions throughout the non-vegetated period. These comprise residual flows in the Karshi and ABMC systems (2.34 km^3), from Karakalpakstan (2.16 km^3), from Khorezm (1.14 km^3) and from all diversions to Turkmenistan (5.25 km^3), altogether accounting for some 10.9 km^3.

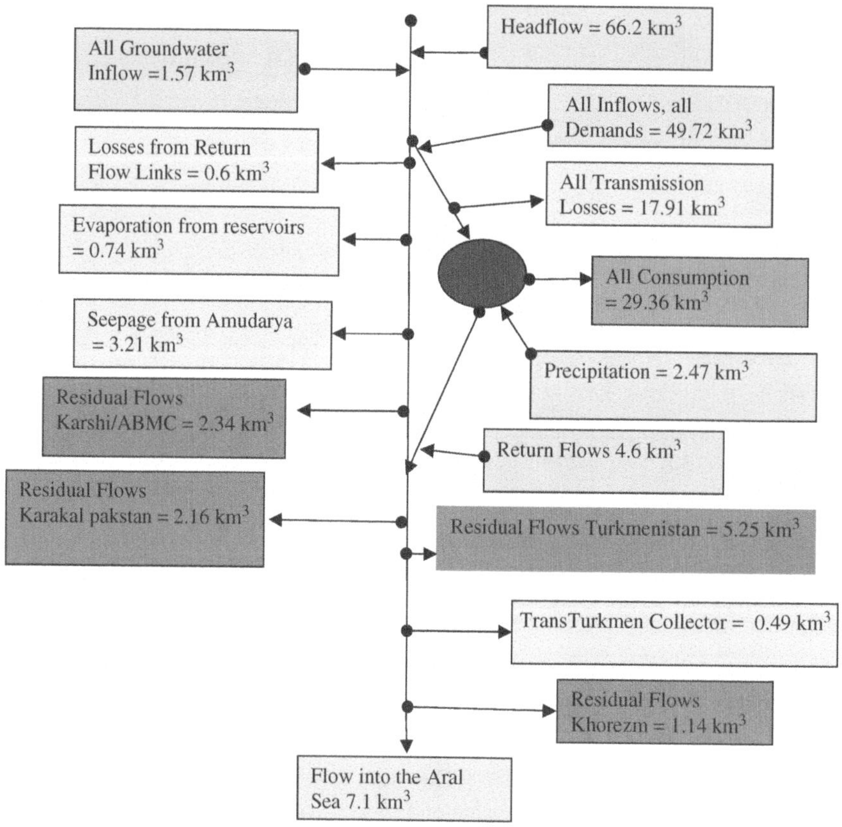

Figure 5.5 AMU-WEAP_100: The Allocation of Water within the Amudarya Basin (based on the annual means from 100 years of values).

The fate of these residuals is unclear. An appreciable amount must be stored in the soil where it will support percolation to the watertable and raise the soil's moisture content; some must inevitably return to the Amu Darya and some clearly must be wasted. Soil moisture stored in such a way provides a valuable resource, which may be used later on in the season by the crops, of particularly relevance during low flow years. *Hence the shortfalls and the implied areas of non-irrigated land are likely to be considerably less than those indicated by the WEAP model.* This, however, is unlikely to affect our overall conclusions that there are major shortfalls or the relative magnitude of differences between the shortfalls predicted for different climate scenario outcomes.

Many irrigated areas are located far from the river, separated both by distance and topography, which hinders or even prevents water from being returned. Bukhara is a case in point: drainage from large parts of it's irrigated area flow to a depression occupying some 590 km^2; this must account for significant quantities of wasted water (evaporation from the depression is estimated conservatively to be 0.4–0.8 km^3 annually). Likewise, the huge irrigated areas in Turkmenistan supplied from the Karakum Canal (Mary and Ahal) appear (from satellite images) to have no obvious direct return flow links to the Amu Darya, similarly for the Dashkauz region. Residual flows from these areas appear to flow to the Sarygamysh Gol lake (area 3,800 km^2) where most must simply be consumed by evaporation (estimated to be of the order of 5 km^3/year).

5.4.1 What happens to the water?

The official data (period 1980–2005) indicates that the mean of the combined annual tributary headflows entering the basin was 66.2 km^3. The official recorded flow entering the Aral Sea (1980–2005) was 8.4 km^3. Based on these values (including 1.6 km^3 from groundwater) some 59.4 km^3 of water should in theory be available for use within the basin. The total consumption of treated water and irrigation appears to account for only 29 km^3, and it therefore follows that some 30 km^3 are not being consumed. The precise figure is the subject of some uncertainty but what is inescapable is that very large volumes of water must either be lost within the irrigated areas, either by seepage, or by disposal to and subsequent evaporation from depressions (see for example the report on the fate of water in the Amu Darya basin by L.K. Smedema, 2000).

The ratio between the estimated losses (17.9+0.6 km^3 = 18.5 km^3, Figure 5.6) and the theoretically available water resource (59.4 km^3) is ~31% (18.5/59.4), although if the residual flows are included this rises to 49% ([18.5+10.9]/59.2).

More realistically, if the demand sites are considered in isolation, the resource available to them was 45.1 km^3 (inflows to – return flows = [49.7 – 4.6]). The ratio between the losses (18.5 km^3 [17.9+0.6] and the available resource is ~41% [18.5/45.1]) which implies that the overall efficiency of water use within the basin is of the order of 60%. This is both realistic as well as being reasonable, although there is clearly considerable scope to make much better use of the excess water.[5]

The mean annual shortfalls for all scenarios considered are depicted in Figure 5.6. The scenarios include: the reference scenario for the existing climate and river flows in 1985. The scenarios for the climate in 2085 apply to the median temperatures, with an upper temperature range value and a lower range value of respectively: 13.6 km^3 and 10.8 km^3. An intermediate value (10.1 km^3) applies to the median climate predicted for 2050.

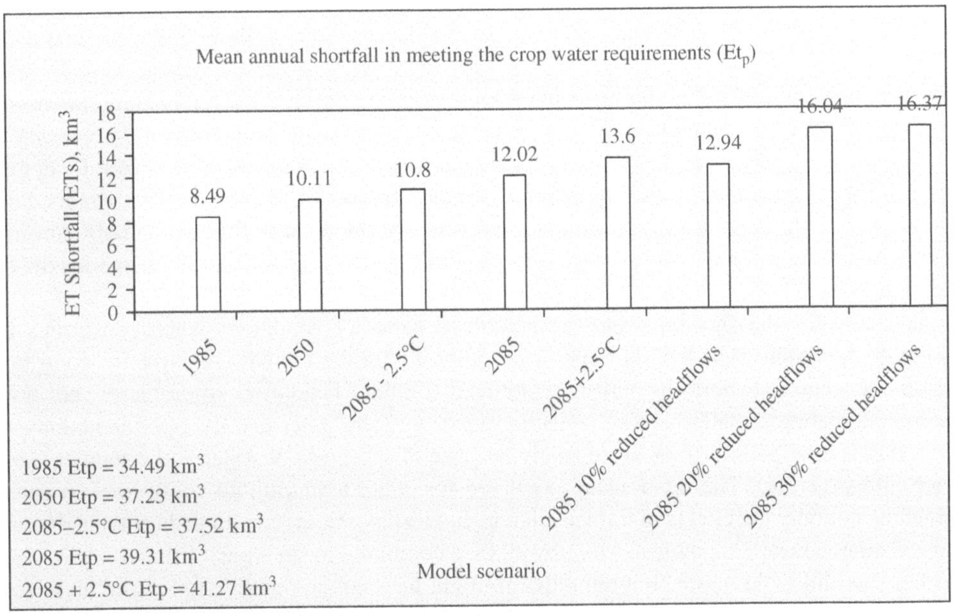

Figure 5.6. The mean annual shortfall of potential evapotranspiration (ETs) for all scenarios included in the model "AMU-WEAP_100". NOTE: Infrastructure Capacity Constraints Apply.

Three more scenarios apply to the median predicted climate for 2085 but then assume a decrease of tributary headflows equivalent to respectively: 10%, 20% and 30%. These indicate that the mean annual shortfalls are likely to be within the range: 12.9–16.4 km^3 (10%–30% flow reduction). Converted into areas, which it is not possible to irrigate, they correspond in the worst case to over 1 million hectares.

5.5 ASSESSMENT OF MODELLING

The results generated by any computer-based model depend upon how well the model has been constructed and how well it represents the system as well as the accuracy of the included data. As stated at the outset, the models are not perfect and would benefit by the inclusion of more accurate data as and when it becomes available. Several sources report discrepancies between the official data for abstraction and irrigated areas. For example, Conrad, 2006, independently recorded abstractions over two seasons in Khorezm and reported that these were some 30% higher than the official values. A related study also reported that the irrigated areas were significantly larger than the official values due to the unapproved expansion of rice production. Despite these reservations, the authors remain confidant that the models are more than capable of simulating the impacts of climate change and increased competition for water resources within the basin.

The schematic, which formed the basis for the structure of the models, was developed in consultation with and using data provided by our partners, and this has been populated using data also provided by them. The version of the model used in development and calibration: AMU-WEAP, relied on 25 years of recorded flow data for the 6 principal

[5] The loss of 40% in delivery and distribution within the irrigated areas implies an efficiency of water use of 60% which is realistic for irrigated areas. It is the fate of the unused excess waters which is of greatest concern.

tributaries of the Amudarya. It became apparent that there were discrepancies in some of the flow data, particularly in the years shortly after independence. The version of the model used to derive results (AMU-WEAP_100) relied on 100 years of flow data generated from the historic record by stochastic analysis. This provided a framework of more statistically more meaningful set of flow data.

Demands within the basin are largely generated by irrigation. The estimates of demand are based on the existing areas and cropping patterns, but in practice the areas devoted to particular crops are not fixed. The model AMU-WEAP_100 does not consider changes of crops, growth of irrigated areas or of populations; since to do so would have led to unrealistic changes over the extended 100 year period considered (e.g. a population growth rate of 2% causes a population of 1 million to rise to over 7 million in the 100 years period, in such a situation treated water consumption could come to dominate demand). Future applications of the model can easily be modified to incorporate established growth factors.

Subject to these reservations, it is considered that estimates of the amount of water needed to satisfy the crop water requirements (ET_p) are realistic. The *actual evapotranspiration* (ET_a) as well as the *shortfalls* [$ET_s = (ET_p - ET_s)$] do, however, depend to a very considerable extent on the quality of the data, the assumptions made and constraints within the model.

Ultimately the main indicator of water shortages have been based on the shortfalls in meeting the needs of irrigation. The relative shortage (ET_s/ET_p) has then been used as a guide to the areas of land which it "*might not*" be possible to irrigate. *These estimates must only be regarded as being indicative guides.* Shortfalls are driven by demands, but some of the water accounted for as "*residuals*" during the non-vegetated season (10.9 km^3) must inevitably contribute to crop water requirements and thus help reduce the shortfalls. Nevertheless, the extent of the potentially non-irrigable areas is very large, and even assuming a considerable margin of error, still pose a very serious challenge for the management of the basin's water resources.

The relationships between the shortfalls and the combined tributary headflows when constraints on conveyance capacity apply (Figure 5.5) indicate that a shortage of capacity within the conveyance systems contributed.

The relationships between the shortfalls and the combined tributary headflows when there were no constraints on capacity (Figure 5.6), arguably, provides a more realistic overview of the water that the riparian nations would consume if it was available. In this case the shortfalls arose solely because of shortages of water at the sources (i.e., the entrances to the conveyance infrastructure). The relationship predicted for 2085, assuming no constraints on capacity, has been extrapolated to include not only the effects of climate change but also possible reductions in tributary headflows, ranging up to 30% below existing values. This indicates that by 2085, a combined mean annual tributary flow of 46.2 km^3 might be associated with a shortfall of 15.4 km^3, implying that it might then not be possible to irrigate up to some 1.3 million ha of the existing 3.4 million ha of irrigated land.

What is quite inescapable is that there is no scope whatsoever for expanding the existing irrigated areas; rather there needs to be a policy of selectively retiring land which for one reason or another (saline land, isolated tracts of land, uneconomic costs of transmission etc.) cannot be or should not be supplied with water.

5.6 REFERENCES

Ahmad, M. and Wasiq, M. (2004) Water resource Development in Northern Afghanistan and its implications for Amu Darya Basin. 29703; World Bank Working Paper No. 36 ISBN: 0-8213-5890-1; ISSN: 1726-5878.

Bedford, D.P. (1997) Climate Sensitivity and Water Management In The Upper Amu Darya Basin, PhD Thesis, the University of Colorado, Boulder, USA (see Chinook University Libraries Catalogue).

Brinkhoff, T. (2007) City Population. Website address: http://www.citypopulation.de/

Bucknall, J., Klytchnikova, I., Lampietti, J., Lundell, M., Scatasta, M. and Thurman, M. (2003) Irrigation in Central Asia, Social, Economic and Environmental Considerations; The World Bank; Website address: http://www.worldbank.org/eca/environment

CAWATERinfo Information Portal of knowledge for Water and Environmental Issues in Central Asia: Water Resources of the Amudarya River Basin (for water consumption). The Portal was created within the framework of the Central Asia Regional Water Information Base (CAREWIB) Project funded by the Swiss Agency for the Development and Cooperation and implemented by SIC ICWC jointly with the UNECE and UNEP/GRID-Arendal. Website address: http://www.cawater-info.net

Chub, V.E. (2000) Climate change and its influence on Natural Resources Potential of the Republic of Uzbekistan. Tashkent, SANIGMI Press, 252 pp. (in Russian).

Chub, V.E., Agaltseva, N. and Myagkov, S. (2000) Climate Change Impact on the Rivers Runoff for the Central Asian Rivers. Central Asian Research Institute (SANIGMI), 72, K. Makhsumova Str., 700052, Tashkent.

Clarke, D., Smith, M. and El-Askari, K. (1998) New software for Crop Water requirements and Irrigation Scheduling. *Journal of the International Commission on Irrigation and Drainage*, **47**(2), 45–58.

Clarke, D., Smith, M. and El-Askari, K. (2000) CROPWAT7 and CROPWAT for Windows. Crop water requirements software package. "Food and Agriculture Organisation of the United Nations, Website address: http://www.fao.org/ag/agl/aglw/cropwat.stm

Conrad, C. (2006) (PhD thesis) Fernerkundungsbasierte Modellierung und hydrologische Messungen zur Analyse und Bewertung der landwirtschaftlichen Wassernutzung in der Region Khorezm (Uzbekistan). This study focuses on the irrigation and drainage systems of Khorezm, located in the lower Amu Darya Basin. Dissertation submitted for the award of: Doctor of the Bavarian Julius-Maximilians-University of Würzburg, Germany.

Dukhovny, V.A., Sorokin, A.G. and Nikulin, A.S. (2002) Amu Darya River Basin: Ways of Probable Conflicts Prevention. Scientific-Information Center of Interstate Water Coordination Commission of Central Asia, email contact dukh@online.ru., Website address: http://afeid.montpellier.cemagref.fr/mpl2003/Conf/DukhovnResu.pdf

ECE (2006) Preliminary assessment of Transboundary Rivers in the Aral Sea Basin and their Major Transboundary Tributaries; Report ECE/MP.WAT/2006/16/Add.4. Website address: http://www.unece.org/env/documents/2006/wat/ece.mp.wat.2006.16.add4.e.pdf

FAO: AQUASTAT (2008) FAO's global information system on water and agriculture, developed by the Land and Water Development Division of FAO (provides statistics concerning agriculture and water use in: Afghanistan, Turkmenistan, Tajikistan and Uzbekistan).Website address: http://www.fao.org/nr/water/aquastat/main/index.stm

FAO (2001) Global Information System and Early Warning System on Food and Agriculture World Food Programme: FAO/WFP CROP AND FOOD SUPPLY ASSESSMENT MISSION TO TAJIKISTAN. Available from Website address: http://www.fao.org/docrep/004/y1534e/y1534e00.htm#P647_36588

FAO (1998) Crop evapotranspiration – Guidelines for computing crop water requirements – FAO Irrigation and drainage paper 56. Authors: Richard G. Allen, Luis S. Pereira, Dirk Raes and Martin Smith. Available from: Food and Agriculture Organisation of the United Nations, Rome, 1998. Website address: http://www.fao.org/docrep/X0490E/x0490e00.htm

Froebrich, J. and Kayumov, O. (2004) Water management aspects of AmuDarya. In *Dying and Dead Seas-Climatic Versus Anthropogenic Causes*. Nihoul JCJ *et al.* (eds.), Vol. 36, NATO Science Series: IV, Earth and Environmental Sciences, Kluwer Academic Publishers, Dordrecht.

Froebrich, J., Bauer, M., Ikramova, M. and Olsson, O. (2007) Water quantity and quality dynamics of the THC – Tuyamuyun Hydroengineering Complex and implications for reservoir operation. *Environmental Science and Pollution Research (ESPR)*, **14**(6), 435–442.

Ivanov, E.G. and Ismailov, G.Kh. (1995) Assessment of natural and anthropogenic losses of the Amu Darya runoff in upstream, middle course, and downstream sections of the river. *Water Resources*, **22**(1), 1–8.

IWMI (International Water Management Institute) On-line Climate Summary Service (2007) Website address: www.iwmi.cgiar.org/WAtlas/AtlasQuery.htm

IWMI Click and Plot section of IDIS also available from Website address: http://dw.iwmi.org/dataplatform/Clicknand Plot.aspx

Khromova, T.E., Osipova, G.B., Tsvetkov, D.G., Dyurgerov, M.B. and Barry, R.G. (2006) Changes in glacier extent in eatsern Pamir, Central Asia, determined from historical data and ASTER imagery. *Remote Sensing of Environment*, **102**, 24–32.

McKinney, D. (2003) Co-operative management of transboundary water resources in central Asia. In *Tracks of Temerlane-Central Asia's Path into the 21st Century*. Burghart, D. and Sabonis-Helf, T. (eds.), National Defense University Press, 2003. Website address:-http://www.ce.utexas.edu/prof/mckinney/papers/aral/CentralAsiaWater-McKinney.pdf. Republic of Tajikistan, Ministry for Nature Protection, Main Administration on Hydrometeorology and Environmental Monitoring, 2002. The First National Communication of the Republic of Tajikistan under the United Nations Framework Convention on Climate Change. Website address: http://www.unfccc.int/resource/docs/natc/tainc1.pdf

SANIIRI-UZVOD ECC Co (2002) JAYHUN: Interstate water resource risk management: Towards a sustainable future for the Aral basin; INCO Contract 516761; Contractor SANIIRI.

Severskiy, I.V. (2004) Water-related problems of Central Asia: Some results of the (GIWA) international water assessment program. *AMBIO: A Journal of the Human Environment*, **33**(1), 52–62.

Sherman, S. and Rafikov, V. (1992) Sedimentation of the Nurek reservoir. *Power Technology and Engineering (formerly Hydrotechnical Construction)*, **25**(10), 668–673.

Smedema, L. (2000) Irrigation-Induced River Salinization: Five Major Irrigated Basins in the Arid Zone. Book: ISBN 92-9090-391-0. Published by the International Water Management Institute, Columbo, Sri Lanka.

Sorokin, A. (2006) Internal Paper for the Kick-off Meeting of Project JAYHUN. Tashkent, Uzbekistan.

Sorokin, A. (2006a) Report on actual data availability and reservoir capacities for the THC Reservoir; JAYHUN Interstate Water Resources Risk Management. "Towards a Sustainable Future for the Aral Basin," Specific Targeted Research Project Contract No. 516761.

Sorokin, A. (2006b) Report on actual data availability and reservoir capacities for the Nurek Reservoir; JAYHUN Interstate Water Resources Risk Management. "Towards a Sustainable Future for the Aral Basin," Specific Targeted Research Project Contract No. 516761.

Sorokin, A. (2008) Assessment of capacity loss dynamics in the reservoirs at the Amudarya Basin: Rogun, Nurek and Tuyamuyun Scientific-Technical Center "Toza Daryo" (Clean River), Uzbekistan; Central Asia Symposium; 24 April, Wageningen, The Netherlands.

Tajikistan Met Service (2002) Tajikistan. Vulnerability of natural resources, economy and public health to climate change. Available from Website address: http://enrin.grida.no/htmls/tadjik/vitalgraphics/eng/html/uyazv.htm

UNECE (2007) United Nations Economic Commission for Europe: Chapter 5 Water Resources Management (Uzbekistan). Available from Website address: http://www.unece.org/env/epr/studies/uzbekistan/chapter05.pdf

UNECE (2004) United Nations Economic Commission for Europe: Chapter 8 Water Resources Management (Tajikistan). Environmental Performance Reviews: Tajikistan. United Nations: New York and Geneva. Website address: http://www.unece.org/env/epr/studies/Tajikistan/chapter08.pdf

UNDP (2003) tapping the POTENTIAL, Improving water management in Tajikistan. National Human Development Report, available from Website address: http://planipolis.iiep.unesco.org/upload/Tajikistan/Tajikistan%20HDR%202003.pdf

Uzbekistan. Microsoft Encarta Online Encyclopedia (2007) Available from Website address: http://encarta.msn.com/encyclopedia_761551989_1/Uzbekistan.html

WEAP Water Evaluation and Planning System Website address: http://www.weap21.org Stockholm Environment Institute_Boston, Tellus Inst itute, 11 Arlington Street, Boston, MA 02166, USA.

Wegerich, K. (2004) Organisational problems of eater fistribution in Khorezm, Uzbekistan. *Water International*, **29**(2), 130–137.

Zonn, I. (2001) 'Water Resources of Northern Afghanistan and their Future Use', Paper presented at workshop on water, climate and development issues in the Amu-Dar'ya basin, Philadelphia, PA.

Chapter 6

Rate of sedimentation of reservoirs, reservoir life and the impact on future water resources

Oliver Olsson, Anatoly Sorokin, Ilkhom Sharipov and Inom Normatov

6.1 SEDIMENTATION PROCESSES IN THE LARGE RESERVOIRS OF THE AMU DARYA RIVER BASIN

Dams and Reservoirs contribute to the improvement of water resources management and play therefore a fundamental role in the water supply worldwide. In semi-arid regions as in Central Asia, water supply for irrigation purposes is the main water sector of consumption. As in this region the water supply relies on storage in surface reservoirs the main influencing parameter is here the reservoir sediment deposition, which affects adversely the water storage.

If the sediment inflow is large relative to the reservoir storage capacity, then the useful life of the reservoir may be very short. Reservoir benefits may be considered sustainable if a storage volume is provided that exceeds the volume of the sediment supply in the tributary watershed. The required sediment storage volume may be included within the reservoir pool or in one or more upstream impoundments.

Sediment depositional patterns are mostly determined by storage operation of reservoirs. This means that reservoirs are operated to maximize water yield, without considering operational changes to limit storage capacity loss due to sedimentation (Basson and Rooseboom, 1997). Considering the limitations of current technology it must be recognized that, with few exceptions, ultimate filling of reservoirs is inevitable and must be taken into account in project planning and design (Vanoni, 2006). Hence, sediments are generally considered an undesirable but unavoidable consequence of water storage. Accepted practice has been to design with the traditional concept of dead storage and operate reservoirs to fill with sediment, generating benefits from remaining storage over a finite period of time (Palmieri *et al.*, 2001). The consequences of sedimentation and project abandonment are left to the future. This future has already arrived for many existing reservoirs and most others will eventually experience a similar fate.

The Amu Darya is the river with the second highest sediment load in the world after the Huang He in China. Gwosdetsky and Michilow (1978) estimated a suspended solid concentration of 3300 mg/l and Letolle (1996) reported concentrations between 1000 and 3500 mg/l. Generally, the main part of the annual sediment load is carried in the summer months (May to September). During the passing of the summer flood, a maximum amount of suspended solids is observed; the minimum however is seen in November and December (Suslov, 1961).

Due to the expansion of the irrigated agricultural area starting in the time of the Soviet Union, the large reservoirs needs to satisfy the generally increased water demand and especially the changed seasonal demand. Therefore it is important to obtain reliable information on present and future water resources and to determine the actual storage of usable water.

Two large dams, the Nurek dam upstream and the Tuyamuyun Hydro-Complex downstream of the Amu Darya River are the dominant regulations in the Amu Darya Basin (Figure 6.1). The Nurek reservoir is located on the Amu Darya tributary Vakhsh about 75 km east of Dushanbe, Tajikistan. The Nurek is a large earth fill dam of 300 m height. Its reservoir, with a design capacity of 10.5 km^3, is the largest reservoir in Tajikistan. It is more than 70 km long, with a surface area over 98 km^2 and a maximum depth of 220 m.

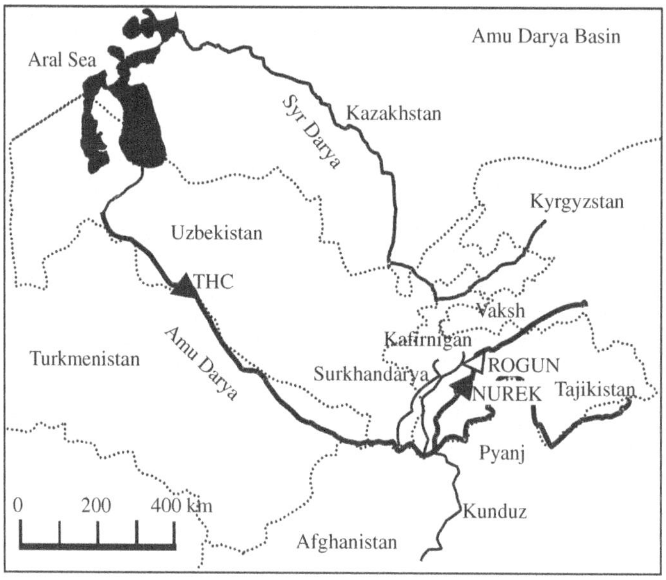

Figure 6.1. The Amu Darya basin with its main tributaries and the dams of Rogun, Nurek and Tuyamuyun Hydro Complex (THC).

The only dam downstream of the Nurek is the Tuyamuyun Hydro Complex (THC), which is located 300 km south of the Aral Sea on the territories of Turkmenistan and Uzbekistan, and is impounding the Aral Sea tributary Amu Darya. The complex consists of more than 30 main hydraulic structures including four interconnected reservoirs: the Channel Reservoir (Amu Darya main stream), the Kaparas reservoir, the Sultansanjar reservoir, and the Koshbulak reservoir with a total design storage capacity of 7.8 km^3. Water from THC is discharged to the lower Amu Darya and to an extensive canal system supplying the regions Khorezm, Karakalpakstan and Tashauz. The provided storage is mainly used for agriculture (around 98%), and also for industry and the drinking water supply (up to 2%).

Especially, the risk of reservoir storage capacity losses at the Tuyamuyun Hydro-Complex (THC) is a serious problem for the future water supply of the lower Amu Darya region downstream of the THC.

6.2 NUREK DAM AND ROGUN DAM PROJECT IN THE VAKHSH RIVER

The Vakhsh is dominated by the Nurek reservoir and the hydraulic potential of the Vakhsh River is mainly used for energy generation but to a small extent also for irrigation purposes. Over 95% of electricity in Tajikistan is generated by hydrolelectricity power plants, particularly Nurek HEP (10.5–11.5 TWh/a) (Schmidt *et al.*, 2006). Originally the Nurek dam was designed and used to increase the water supply during the irrigation season. Today the reservoir is also used to release water for hydropower production during the winter months (Wegerich *et al.*, 2007).

The Nurek reservoir bottom has a design level of around 680/690 m above sea level (a.s.l.), and the normal pool level (NPL) elevation is 910m (a.s.l.). The operation of the Nurek reservoir is characterized by water level variations between the maximum operating level of 910 m (a.s.l.) and minimum operating level of 857 m (a.s.l.). Within this range, the active regulation storage of the conservation pool is 4.5 km^3, providing seasonal stream flow regulation of the Vakhsh River (Sherman and Rafikov, 1992), while in total the inactive/dead storage between the water levels of 680 m and 857 m (a.s.l.) amounts to 4.0 km^3, according to the design capacity. The hydropower capacity is 3000 MW and the long-term average annual hydropower production is 11.2TWh.

Nurek's current mode of operation demonstrates its dual role of water control to support downstream irrigated agriculture and to produce energy. Figure 6.2 shows daily values for water levels, inflow and outflow in 2003 and 2004. The minimum inflow, 57 m^3/s in 2003 and 64 m^3/s in 2004, occurs during February, with a rapid increase of inflow beginning in March. There is a characteristic sequence of flood events leading to a continuous increase of the average flow. The water level variation is characterized by a continuous decrease during the winter and spring months until the minimum level is reached in May (17 May 2003: 858 m; 6 May 2004: 856 m). Directly after reaching the minimum level, refilling of the reservoir commences and the maximum water level of 910.5 m is reached in autumn (19 August 2003 and 11 September 2004). Subsequently, there is a continuous decrease until May of the following year. This indicates that the Nurek dam is currently prior utilized to regulate the flow of the Vakhsh River for irrigated agriculture, in downstream Turkmenistan and Uzbekistan (Figure 6.2).

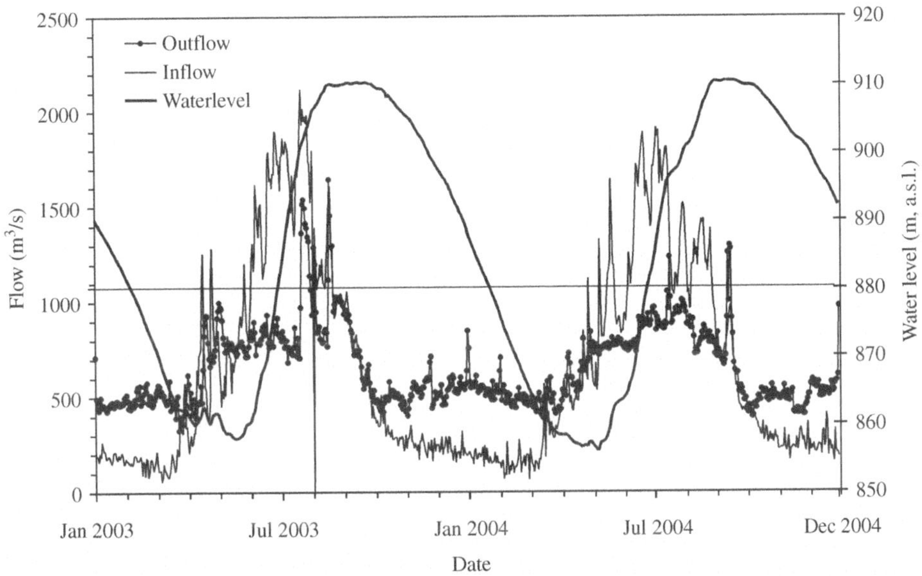

Figure 6.2. Daily water level variations, inflow and outflow for Nurek reservoir for 2003 and 2004 (Ministry of Energy, Tajikistan, 2004).

To increase its energy output, Tajikistan is planning to recommence the construction of the Rogun reservoir (3,600 MW), 100 km northeast of the Tajik capital – a project that was started during the Soviet period but stopped with the Tajik civil war in 1991. The Uzbek government is highly critical of the Rogun dam (Spoor and Krutov, 2003), because it would "put it [Tajikistan] firmly in control of the river" (ICG, 2002). However, neither Spoor and Krutov (2003) nor the ICG (2002) report distinguish between different stages of the Rogun construction. It is not clear whether Uzbekistan is opposed to the construction in general or to a particular stage of construction.

The World Bank, and more recently the feasibility study of the construction company Lahmeyer International (Schmidt *et al.*, 2006) did this and distinguished between different stages of dam construction. In Stage I, Rogun is supposed to provide an annual energy output of 5.6TWh. For this purpose the height should be 225 m, with a total reservoir volume of 2.78 km^3, live storage of 1.92 km^3 and installed capacity of 1,000 MW. In Stage II, the dam height should be raised to 285 m (reservoir volume 6.78 km^3 and live storage 3.98 km^3) and in Stage III to 335 m (reservoir volume 13.3 km^3 and live storage 10.3 km^3).

6.2.1 Sediment dynamics in the Vakhsh river

The Vakhsh river, also known as Kyzyl-Suu in Kyrgyzstan and as the Surkhob in north-central Tajikistan, originates in the alpine regions of the Pamir Alai in the south-east of Kyrgyzstan, where parts of the Abramov glacier and the Fedchenko glacier contribute to the runoff generation. The Vakhsh River flows from north-central Kyrgyzstan to the south-west of Tajikistan. Until its confluence with the Pyanj, the Vakhsh has a length of 524 km, and a catchment area in Tajikistan of 31.2 thousand km^2. The largest tributaries of the Vakhsh are the Miksu and the Obihingou (UN, 2004). The average annual runoff of the Vakhsh River at the site of the Nurek dam is equal to 20 km^3, there is a 10% probability of a 23 km^3 runoff, and a 90% probability that runoff will not exceed 16.6 km^3. Giese *et al.* (2004) report an annual mean discharge of 20.0 km^3/a for the Vakhsh and a mean discharge of 34.3 km^3/a for the Pyanj. The Vakhsh contributes only 25% to the total mean discharge of the Amu Darya (79.3 km^3/a).

The Vakhsh River is characterised by frequent land slides and avalanches. In addition, the very steep mean slope of the Vakhsh results in a very high transport capacity. The intensive use of fuel wood during the civil war decreased the forest areas and volumes and led to mountain slope deforestation. Consequently, there is now an increased risk of mudflows and soil erosion (Makhmadaliev *et al.*, 2003). It is unlikely that the need for fuel wood utilization will change, because of the high level of poverty.

The normal annual sediment load discharge of the Vakhsh river to the Nurek reservoir in the design was estimated to be 88.7 million m^3/a, whereas the maximum sediment load of 4400 kg/s was reached at a discharge of 782 m^3/s and the minimum sediment load of 1470 kg/s at a discharge of 487 m^3/s (Table 6.1). The sediment composition of the reservoir inflow was described with a particle content of 37.5% sand, 20.4% silt, and 42.1% clay (Table 6.2).

The proportional distribution of the sediment load on the seasonal discharge is shown in Table 6.3. It describes the main proportion of sediments with 62% on the annual sediment load for the time period from July to September, which is also the time period of the maximum proportion (49%) on the annual river discharge.

Table 6.1. Annual average sediment load of the Vakhsh river as inflow reference to Nurek reservoir.

	Discharge, m^3/s	Sediment-load, 1000 kg/s
Max rate	782	4.4
Min rate	487	1.47
Average	640	2.8

Table 6.2. Typical sediment composition for Vakhsh river as inflow reference to Nurek reservoir.

Material	Diameter, particle mm	Content by weight, %
Clay	less 0.01	42.1
Silt	0.05–0.01	20.4
Fine sand	0.25–0.05	34.2
Coarse sand	more 0.25	3.3

Table 6.3. Seasonal discharge and the proportion on the annual sediment transport for Vakhsh river.

Period	Discharge, %	Sediment load, %
October–February	15	1
March–June	36	37
July–September	49	62

6.2.2 Assessment of reservoir storage losses of the Nurek Dam reservoir

A recent World Bank report (2005) states that, during the last 25 years, roughly 50 meters of the 300 meters have been lost due to silt and that Nurek will be able to operate even without silt removal for at least another 30 years. An earlier World Bank report (2003) had argued, "The Nurek reservoir's capacity has been reduced by 67% over the past 26 years".

Figure 6.3 shows the original design capacity (bold line) of the Nurek reservoir for different water levels. Sherman and Rafikov (1992) argued that the geometry of the body of sediments according to the 1989 survey (grey dashed line) shows that the volume of deposits after the 18 years of operation is 1.84 km^3. In October 1989 8.66 km^3 remained unsilted. The volume of deposits within the dead storage below elevation 857 m (a.s.l.) is about 700 million m^3, which means that the useful volume has been silted by 1.1 km^3.

Since the initiation of impoundment of the Vahksh in 1972, at the normal pool level of 910 m (a.s.l.), storage capacity has been reduced by 17% from 10.5 km^3 to 8.7 km^3. This is an average of slightly over 62 million m^3 per year.

In comparison to the data from 1989, the processed data (2001) obtained from the WB/GEF project – security of the dams and reservoirs indicates no further capacity losses in the total storage volume. However, the recent dataset presents discrepancies within the sedimentation processes of the usable and dead storage volume. The result on the hypothetical capacity losses shows a regeneration of the usable storage volume by reduced losses from 1.12 km^3 (1989) to 0.24 km^3 (2001). Increased sedimentation in the dead storage was present and reduced the capacity from 0.7 km^3 (1989) up to 1.7 km^3 (2001).

Therefore, it is questionable whether the World Bank estimate of the lifespan of the reservoir can be accepted unreservedly. In any case, Nurek will slowly lose its capacity to store the water from the summer floods, and also its importance for season water utilization in the downstream regions.

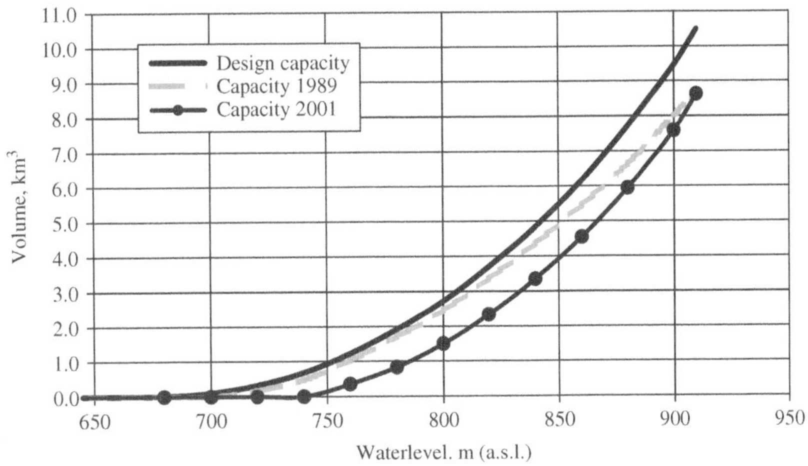

Figure 6.3. Volume elevation rating curves of Nurek reservoir.

Nurek-Rogun cascade

Utilizing the Nurek in combination with the Rogun dam could be a win-win solution for downstream and upstream countries. It would guarantee Tajikistan's energy requirements and at the same time, it would enable Tajikistan to continue to facilitate the agricultural production of downstream riparian states. Furthermore, Stage II of the Rogun dam could be seen as a safety measure for times of water scarcity or drought. During a period of drought, additional releases from Rogun could increase the available water for downstream urban and agricultural users. It appears that, with the additional dam, the benefits to all the riparian states could be increased.

Rogun will be not only of high interest for the power production, but has also significant positive effects on siltation and storage capacity losses of the existing downstream reservoirs. Due to sediments held back by Rogun dam, the life-time of Nurek, which is located 70 km downstream of Rogun, will be substantially prolonged.

Prediction of reservoir storage capacity losses for Nurek/Rogun cascade

For the prediction of future reservoir storage capacity losses the calculation of sedimentation rates was executed. Therefore two calculation scenarios for the sedimentation of the Nurek and Rogun reservoirs are proposed:

Variant 1 – Nurek without Rogun reservoir, and

Variant 2 – joint work of Rogun and Nurek reservoirs, if Rogun reservoir influences the sedimentation in the Nurek reservoir.

A balance method was applied for the calculation of sedimentation rates for the large Nurek and Rogun reservoirs. This method based on Brune's curves, which are recommended for an estimation of sedimentation of deep reservoirs by Committee on Sedimentation of Reservoirs of International Commission on Large Dams and provides acceptable estimates for large reservoirs.

The study used for the calculation of the both Variants two hydrological scenarios for the available reservoir inflow. Both scenarios were estimated using the results of the climate change assessment and its impact on the upstream run off formation for the next 50 years (Jayhun reports D.3.1 and D.3.2). The scenario (I.) indicated decreasing flow of 3 km^3 to 6 km^3, 5–10%, for the Amu Darya from 1990–2050. If climate change will have more impact on the river flow, the flow is decreasing up to 10–15% until 2050 (Scenario II.). Scenario (I.) indicates the maximal inflow of water and sediment into reservoirs and accordingly maximal losses of water storage in reservoirs. Therefore this scenario (as "worse" case) was accepted as a basis for an estimation of inflow to the Rogun (Nurek) reservoir.

6.3 THE LARGE RESERVOIRS OF THE TUYAMUYUN HYDRO-COMPLEX (THC)

Before 1979, there were only some natural depressions in the area of the current Tuyamuyun Hydro Complex (THC). During the period of 1981 to 1983, the construction was completed and the main stream of the Amu Darya River was impounded at the section of the existing hydro-unit (dam structure). The THC multi-reservoir system consists of four reservoirs, the Channel Reservoir that impounds the natural riverbed, Kaparas, Sultansandjar, and Koshbulak, which constructed as off-stream reservoirs. From Channel Reservoir, the water is either channelled into Kaparas or Sultansanjar reservoir, discharged to the downstream part of the river, from where it flows down the river or enters the irrigation canals. In Channel Reservoir, a minimum water level is required for enabling inflow into Kaparas or

Table 6.4. Predicted reservoir storage capacity losses for the Nurek and Rogun dams.

Year	Variant 1 Nurek, km^3	Variant 2 Nurek, km^3	Rogun, km^3	Sum, km^3
Scenario 1 – with influence of climate change				
2010	2.7	2.7	0.0	2.7
2020	3.1	2.8	0.4	3.2
2030	3.6	2.9	1.1	4.0
2040	4.1	3.0	1.6	4.6
2050	4.4	3.1	2.1	5.2
Scenario 2 – without influence of climate change				
2010	2.7	2.7	0.0	2.7
2020	3.2	2.9	0.4	3.3
2030	3.8	3.0	1.2	4.2
2040	4.3	3.1	1.8	4.9
2050	4.7	3.2	2.5	5.7

Note: Variant 1 – Nurek reservoir, without Rogun reservoir; Variant 2 – Nurek and Rogun reservoirs.

Sultansanjar reservoir, because the water exchange between the reservoirs is based on free surface slope. Separation gates allow an individual management of the single reservoirs.

The main THC dam is a concrete overflow dam with a height of 25 m, a total length of 141 m and a width of 41 m. Its main reservoir the Channel Reservoir, with a design capacity of 2.3 km^3, is the largest reservoir of the THC (Table 6.5). It is more than 102 km long, with a surface area of over 303 km^2 and a maximum depth of 20 m. The reservoir bottom has a design level of around 110 m above sea level (a.s.l.), and the normal pool level (NPL) elevation is 130 m (a.s.l.).

Table 6.5. THC reservoirs, characterised by normal pool level (NPL), dead volume level (DVL), total and active design storage capacities and the area of reservoir surface.

	Reservoir	NPL/DVL, m (a.s.l.)	Total/Active storage capacity, km^3	Area of surface at NPL, km^2
1	Channel Reservoir	130/120	2.34/2.1	303
2	Kaparas	130/117	0.96/0.61	70
3	Sultansanjar	130/116	2.69/1.63	149
4	Koshbulak	130/120	1.81/1.02	128
	Total		7.8/5.36	650

6.3.1 Sediment dynamics in the lower Amu Darya river

For the lower Amu Darya monthly average sediment loads at the Darganata gauge above THC is given in Figure 6.4 below.

The suspended solid and seasonal discharge at the upstream Darganata station is shown in Figure 6.4. For the presented high discharges of 1992, which had a maximum discharge of 4000 m^3/s in July, the graph show very high concentrations of suspended solid with more than 4500 mg/l, during the first flood events of the summer period and low concentrations (around 500 mg/l) during the winter period. However, in the low flow year, 1986, which had a maximum discharge of 1800 m^3/s for July, sediment concentration only rose to 1800 mg/l in May and reached a maximum concentration of 3000 mg/l in January.

The suspended load of Amu Darya River at Darganata arises from upstream bed and bank erosion (Suslov, 1961). Using the Russian soil classification system Sorokin (2008) describes the suspended sediment composition of the THC reservoir inflow with particle content of 15.5% sand, 22% silt, and 62.5% clay (Table 6.6).

For the discharge of the lower Amu Darya River to the THC reservoir at Darganata station from 1981 to 1996, the maximum concentration of suspended solids of 4800 mg/l was measured at a discharge of 1477 m3/s and the minimum concentration of suspended solids of 300 mg/l at a discharge of 710 m3/s (Figure 6.5). The results show an increased trend in distribution of sediment concentration at the Darganata discharge with high sediment concentrations at high Amu Darya discharges.

The THC and the various water distribution structures in the river significantly influence the sediment regime of the lower Amu Darya. The decrease of transported material clearly shows the existence of sedimentation processes and it is

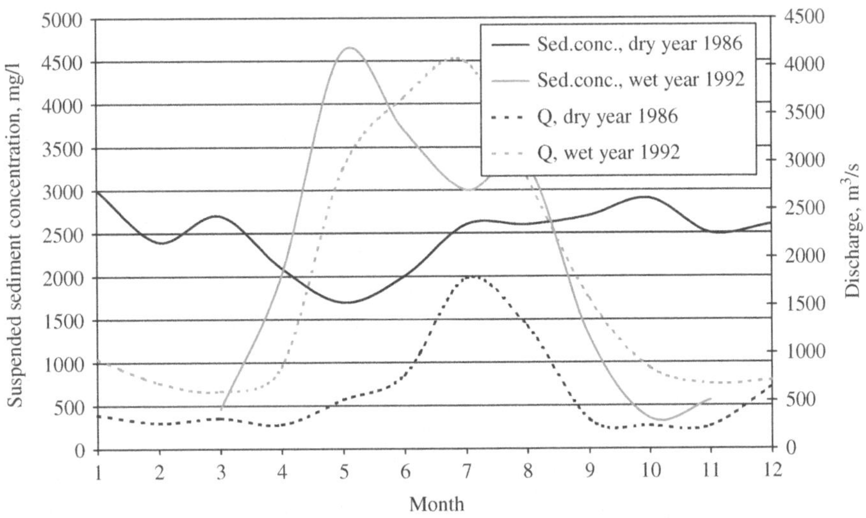

Figure 6.4. Monthly averaged suspended sediment concentration (mg/l) and monthly discharge (m^3/s) at Darganata station, for the wet year 1986 and dry year 1992.

Table 6.6. Typical sediment composition for the lower Amu Darya River (inflow to the reservoir system), Russian soil classification system.

Material	Diameter, mm	%
Clay	less 0.001	62.5
Silt	0.05–0.001	22.0
Fine sand	0.25–0.05	15.0
Coarse sand	more 0.25	0.5

obvious that THC is acting as 'sediment trap'. The data sampled downstream from the complex illustrates (Fig. 6.5), that at the station Tuyamuyun (8 km downstream from THC) the concentration of suspended solids in natural river water ranges between 14 mg/l and 1200 mg/l. Compared to the records before THC it is obvious that a huge amount of sediments is stored in the reservoirs.

6.3.2 Assessment of reservoir storage losses of the Tuyamuyun Hydro-Complex

A serious problem for the multi-reservoir system THC is the loss of storage volume due to sedimentation. Especially in the Channel Reservoir as an in-stream reservoir, occurring sedimentation processes are strongly influenced by the seasonal variation of the Amu Darya runoff. The main proportion of inflowing suspended solids from the river is silted in this reservoir. Whereas, for the off-stream reservoirs Kaparas, Sultansanjar and Koshbulak comparably low storage capacity losses are caused by sedimentation with only the very fine fraction (fine silt and clay – smaller than 0.005 mm) is deposited in these three impoundments.

Historic data on reservoir capacity and sedimentation processes has been reviewed and documented for the Tuyamuyun Hydrocomplex. Additionally, field campaigns at the Tuyamuyun reservoirs were executed to investigate the actual bathymetry. The current available storage capacities of the THC reservoirs were determined at the Channel Reservoir (2005, 2008), Kaparas (2005), Sultansanjar and Koshbulak (2008).

Utilizing GPS/sonar the CEEDUCER hydrographic survey system carried out the measurement works. CEEDUCER is a new portable system providing the highest standards of survey and accuracy. Depth range is from shallow water (30 cm) to 100 m and the velocity of measurements is 6 measurements per second. The system is convenient for work at not big platforms and boats. It has 12 channel GPS receiver of a high accuracy. High quality data on total volumes, water levels and reservoir areas were obtained from the bathymetric survey and were used for the post-processing of measured profiles in GIS software. Detailed GIS maps are now available for the current reservoir status.

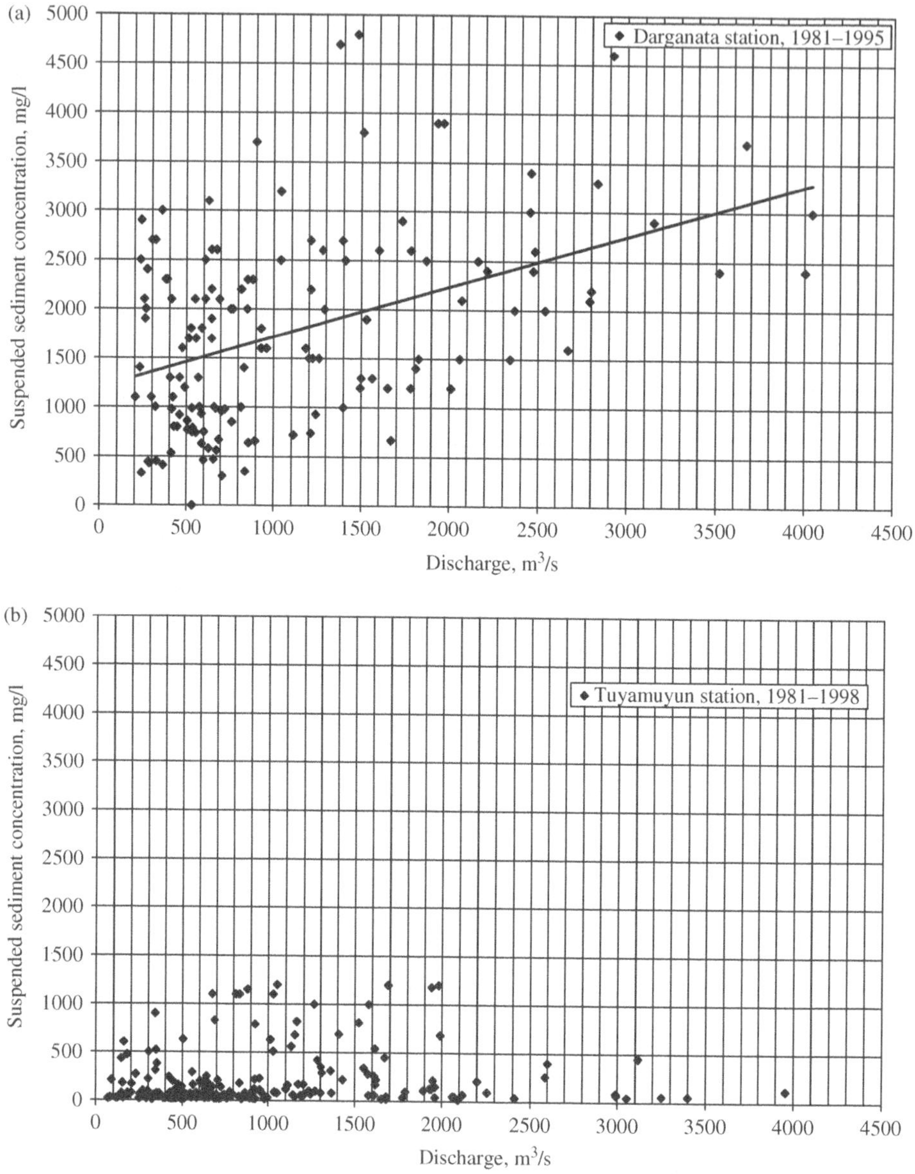

Figure 6.5. Distribution of suspended sediment concentration (mg/l) on the Amu Darya river discharge: (a) Darganata station, 1981–1995, (b) Tuyamuyun station, 1981–1998.

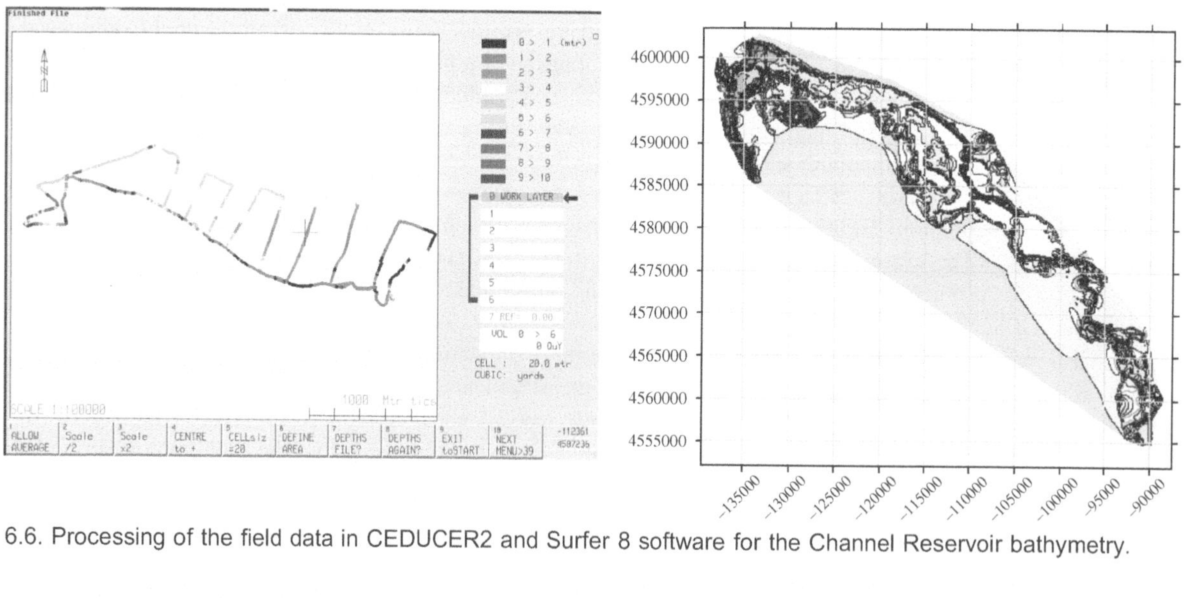

Figure 6.6. Processing of the field data in CEDUCER2 and Surfer 8 software for the Channel Reservoir bathymetry.

Channel reservoir

Since the beginning of the reservoir filling in 1981 the Central Asian Scientific Research Institute of Irrigation (SANIIRI) has carried out yearly field investigations of the Channel Reservoir. The most recent information on Channel Reservoir storage capacity was obtained from the bathymetric survey in 2005 and 2008.

These studies showed a reduction of reservoir storage capacity from 2340 million m^3 in 1983 to 1287 million m^3 in 2005 at the maximum water level of 130 m a.s.l., since the initiation of impounding the Amu Darya River in 1983 (Figure 6.7). This is comparably to a storage capacity loss, of 1053 million m^3 or 45%, and an average of slightly over 48 million m^3 per year. For illustration, Figure 6.7 shows the original design capacity (dashed line) compared to the storage capacity in 2005 (bold line) for Channel Reservoir related to different water levels. The usable storage volume was reduced from 2070 million m^3 to 1251 million m^3, which equals to a siltation of 819 million m^3. The volume of deposits within the dead storage below elevation of 120 m (a.s.l.) is about 234 million m^3 and describes a storage reduction from 270 million m^3 to 36 million m^3.

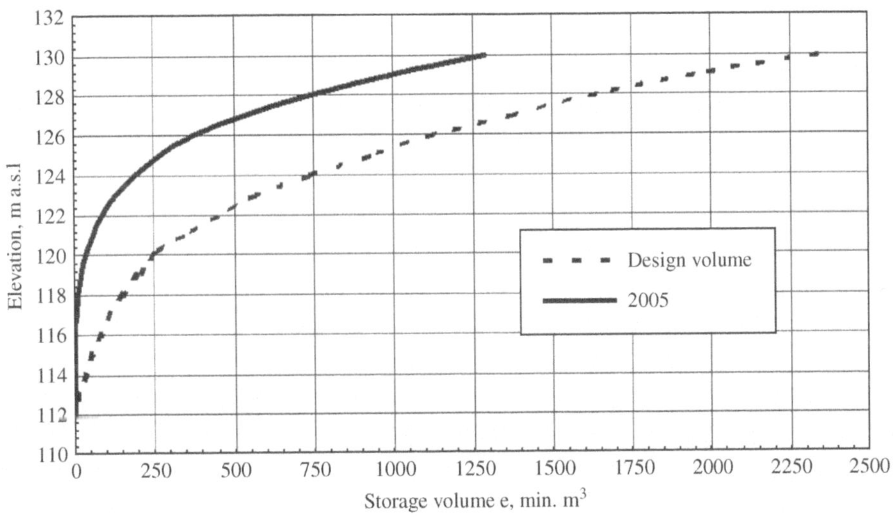

Figure 6.7. Volume elevation rating curves Channel Reservoir (THC), design 1981 and 2005.

The annual variation of storage capacity of the Channel Reservoir and accumulation of sediments, for 1981–2005 is given in Figure 6.8.

It can be observed that the rate of storage loss is initially high in the beginning of the THC operation (1981–1985) and declines during the operation period of 24 years (Figure 6.8). Morrison and Fan (1998) give reason for that by the gradual compaction of cohesive sediment: the volume occupied by cohesive sediment during the first year after deposition is greater than the volume it will occupy after many years of compaction. The lowest sediment deposition can be identified during 2002–2005.

The velocity and the turbulences vary according to water level of Channel Reservoir, which ranges between 112 m (a.s.l.) and 130 m (a.s.l.). If the reservoir is filled, flow velocities are low and silting processes occur. If the reservoir is not filled completely, higher flow velocities occur resulting in higher turbulences and lower sedimentation in a central channel. These characteristics can be observed form the provided annual bathymetric information (SANIIRI) for the high and low water years.

Kaparas reservoir

Inflowing water to Kaparas reservoir has already passed through the Channel Reservoir and therefore has been reduced in velocity and sediment content. Hence, the sedimentation volume in Kaparas reservoir is significantly smaller than in Channel Reservoir. Kaparas reservoir has operated since 1985 and has a design volume of 960 million m^3 at a maximum water level of 130 m (a.s.l.).

The Figure 6.9 presents the most recent bathymetric information on the Kaparas reservoir storage obtained from a bathymetric survey in 2005.

The results for Kaparas describe a maximum total storage volume of 940.85 million m^3 and a water surface area of 65.5 km^2. The obtained information show further an occurring storage capacity loss of 19.15 million m^3 since 1985 with an annual average storage capacity loss of slightly over 960 000 m^3 only about 0.1% per year.

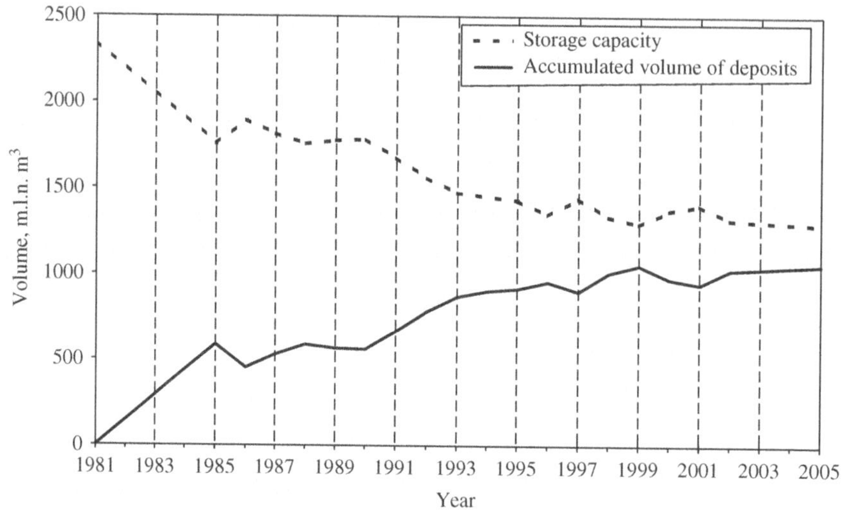

Figure 6.8. Decrease in storage capacity and accumulation of sediments, channel reservoir, 1981–2005.

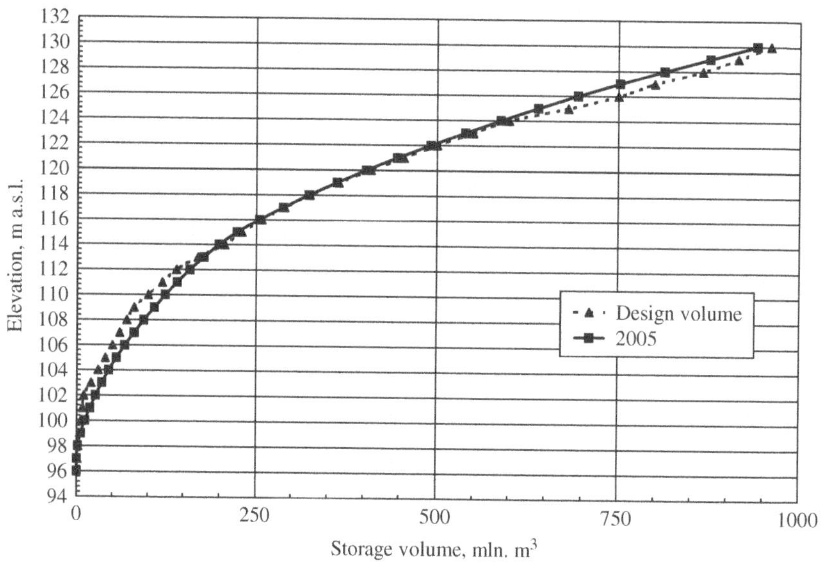

Figure 6.9. Volume elevation rating curves Kaparas reservoir (THC), design 1981 and 2005.

In the year of operation-start in 1981 the dead storage volume of Kaparas was about 320 million m^3 and provides sufficient volume for sedimentation. Storage capacity losses of Kaparas due to sedimentation are therefore more a minor problem.

The THC reservoirs Sultansanjar and Koshbulak were also investigated within this study (Jayhun report D.2.2) and the recently obtained bathymetric information are summarised in Table 6.7. Both Sultansanjar and Koshbulak are off-stream reservoirs and their inflowing water has also already passed Channel Reservoir and has been reduced in velocity and sediment content and they have similar annual storage loss rates like Kaparas.

Table 6.7. Characteristics of the Kaparas, Sultansanjar and Koshbulak reservoirs (field campaigns 2005, 2008, BMC).

Characteristics	Sultansanjar (2008)	Kashbulak (2008)	Kaparas (2005)
Capacity (km^3)	2.65	1.47	0.94
Area (km^2)	147	110	66
Sediment deposition (km^3)	0.04	0.34	0.02

6.4 CONCLUSIONS

During the first 18 years of Nurek reservoir operation since 1972, practically the entire solid runoff of Vakhsh has accumulated in this reservoir with about 0.1 km^3/a, indicating a loss of storage of about 1% per year. This would imply an estimated loss by 2009 of about 35% of its design volume. In spite of this, Nurek dam could be able to operate even without silt removal for at least another 60 years.

Since there is no significant lateral inflow between Rogun and Nurek reservoirs, the same average annual sediment deposition rate may also be assumed for the Rogun reservoir. Thus, after the commissioning of the Rogun dam, sedimentation at Nurek will significantly reduce, thereby increasing the remaining economic life-time of Nurek reservoir by at least 15 years. However, the work commenced on the Rogun Dam (in 1976), the construction stopped in 1993 when a large flood destroyed a cofferdam, which had reached a height of 60 m. Most of the dam and the impounded sediment added to the sediment load of the Nurek reservoir.

The THC significantly influences the sediment regime of the lower Amu Darya by acting as a sediment trap. The results illustrate that a huge amount of sediments is stored at the THC reservoirs. Initially, THC had a total storage capacity of 7.8 km^3 but due to siltation, by 2005 the total storage was reduced to 6.7 km^3.

Since the initiation of impoundment of the Amu Darya in 1983, Channel Reservoirs storage capacity has been reduced by 45%. This is an average of slightly over 2% to 3% per year. In consideration of the results for the potential reservoir capacity losses it can be argued, that Channel Reservoir with a currently available total storage volume of 1,287 million m^3, will be able to operate without silt removal for at least another 25 years.

The lifespan of the other reservoirs, especially Kaparas with around 350 years, will be significantly longer. This can be concluded by the available outcomes from the storage capacity assessment of the off-stream reservoirs, which results in an annual storage capacity loss of 0.1%.

It is clear that due to the high silt load all of the on-stream reservoirs are silting up at an alarming rate. Siltation will significantly adversely affect the ability of the 35+ reservoirs in the Amu Darya basin to regulate seasonal demand for both irrigation and municipal water supply.

It is of critical importance for the Amu Darya Basin to use existing infrastructure in a better way and adapt the water allocation accordingly, especially during dry years. Hereby, the large dams and reservoirs of the region play an essential role as they provide the potential to act as an instrument for adapting to ongoing water shortage caused by global change.

In particular the downstream reservoir complex THC offers more opportunities in this context than the other dams, Nurek and Rogun. This is given by the fact that on the one hand the Rogun reservoir isn't finished so far without any prospect of completion and on the other hand the Vaksh River represents not the main water source for the downstream area of the Amu Darya.

6.5 REFERENCES

Basson, G.R. and Rooseboom, A. (1997) Dealing with Reservoir Sedimentation. Water Research Commission Report No. TT 91/97.

Giese, E., Sehring, J. and Trouchine, A. (2004) Zwischenstaatliche Wassernutzungskonflikte in Zentralasien. Zentrum für internationale Entwicklungs- und Umweltforschung (ZEU), Giessen, Discussion Paper 18, http://www.uni-giessen.de/zeu/Papers/DiscPap%2318.pdf. Accessed May, 2006.

Gwosdetsky, N.A. and Michilow, N.E. (1978) Physical Geography of USSR. Moscow. 3rd edition, p. 28 (in Russian).

ICG (International Crisis Group) (2002) Central Asia: Water and Conflict, Asia Report No. 34, Osh, Brussels.

Jayhun report D.2.2 (2007) Report on actual data availability and reservoir capacities for Nurek reservoir. JAYHUN Interstate Water Resources Risk Management. "Towards a Sustainable Future for the Aral Basin," Specific Targeted Research Project Contract No. 516761.

Jayhun reports D.3.1 (2007) Report on present runoff formation. JAYHUN Interstate Water Resources Risk Management. "Towards a Sustainable Future for the Aral Basin," Specific Targeted Research Project Contract No. 516761.

Jayhun report D.3.2 (2009) Report on future changes in available water resources JAYHUN Interstate Water Resources Risk Management. "Towards a Sustainable Future for the Aral Basin," Specific Targeted Research Project Contract No. 516761.

Létolle, R. and Mainguet, M. (1996) *Der Aralsee – eine ökologische Katastrophe*. Springer Verlag, Berlin, Heidelberg, New York.

Makhmadaliev, B.U., Novikov, V.V., Kayumov, A.K. (eds.) (2003) The first national communication of the Republic of Tajikistan under the United Nations framework convention on climate change, 2nd Phase, Ministry for Nature Protection of the Republic of Tajikistan, http://unfccc.int/resource/docs/natc/tajnc1add.pdf. Accessed May, 2006.

Morris, G. and Fan, J. (1998) Reservoir Sedimentation Handbook. McGraw-Hill, New York.

Palmieri, A., Shah, F. and Dinar, A. (2001) Economics of reservoir sedimentation and sustainable management of dams. *Journal of Environmental Management*, **61**(2), 149–163.

Schmidt, R., Zambaga-Schulz, S. and Seibitz, M. (2006) Bankable Feasibility Study for Rogun HEP Stage 1 construction completion in Tajikistan. In *Dams and Reservoirs, Societies and Environment in the 21st Century*. Berga, L., Buil, J.M., Bofill, E., De Cea, J.C., Garcia Perez, J.A., Mañueco, G., Polimon, J., Soriano, A., Yagüe, J. (eds.). Taylor and Francis, London, pp. 405–413.

Sherman, S. and Rafikov, V. (1992) Sedimentation of the Nurek reservoir. *Power Technology and Engineering (formerly Hydrotechnical Construction)*, **25**(10), 668–673.

Spoor, M. and Krutov, A. (2003) The 'Power of Water' in a divided Central Asia. *Perspectives on Global Development and Technology*, **2**(3–4), 593–614.

Suslov, S.P. (1961) Physical geography of AsiaticRussia. Freeman and Co. San Francisco and London.

UN (United Nations Economic Commission for Europe) (2004) Chapter 8: Water Resources Management. *Environmental Performance Reviews: Tajikistan*. United Nations: New York and Geneva.

Vanoni, V.A. (ed.) (2006) Sedimentation Engineering. ASCE Manuals and Reports on Engineering Practice No. 54.

Wegerich, K., Olsson, O. and Froebrich, J. (2007) Reliving the past in a changed environment: Hydropower ambitions, opportunities and constraints in Tajikistan. *Energy Policy*, **35**, 3815–3825.

World Bank (2003) Memorandum of the President of the international development association to executive directors on a country assistance strategy for the Republic of Tajikistan, Report No. 25329-TJ.

World Bank (2005) Tajikistan: Trade Diagnostic Study, Report No. 32603-TJ.

Chapter 7

Combined reservoir management of water and sediments for the channel reservoir at the lower Amu Darya River

Oliver Olsson, Alisher Khodjiev and Malika Ikramova

7.1 RESERVOIR SEDIMENT MANAGEMENT

Combined reservoir management of sediments as well as water should be considered as an effective measure to improve the sustainable long-term use and to contribute towards a safe water supply. Several authors have emphasised that the concept of a reservoir life limited by sedimentation has to be replaced by a concept of managing both water and sediment to sustain reservoir function (Basson and Rooseboom, 1997; Morris and Fan, 1998).

In semi-arid regions as in Central Asia, water supply for drinking and irrigation purposes depends strongly on storage in surface reservoirs and sediment deposition in these reservoirs adversely affects the water storage capacity.

The focus of this study was on the Tuyamuyun Hydro-Complex (THC) impounding the lower Amu Darya River and its main Channel Reservoir. This in stream reservoir was characterised with high sedimentation dynamics and storage losses during the last 20 years of operation (Chapter 6).

The dams of the THC play a critical role in the integrated management of ground- and surface waters downstream. Both, seasonal water availability, water quality and sediment dynamics depend on the operation and management rules of the THC. Half of the drinking water supply in the lower Amu Darya region comes from the Tuyamuyun reservoirs. In terms of water allocation a major interstate conflict spot can be identified between Turkmenistan and Uzbekistan at the lower Amu Darya because of the THC. In comparison to the past Soviet system, especially the independency of Uzbekistan and Turkmenistan results in a much more complicated situation. Both countries have large irrigated areas and the question of regulating the multi-reservoir system is rather critical, and of strategic and political importance. The main part of the reservoir system is located on the territory of Turkmenistan, except the main dam with its operation unit and the main reservoir, which forms the border between Uzbekistan and Turkmenistan.

This study aims to identify applicable reservoir operations for the THC reservoirs with regard to a combined water and sediment management. For this purpose a simple water budget model (THC-model) and a 3D reservoir sedimentation model were used in conjunctive. Reservoir management scenarios for dry, median and wet years determined by the THC-model were used for the simulation of sediment dynamics in the reservoirs. The simulation results for the adapted management scenarios were compared with simulation results obtained from current practiced THC management scenarios.

The impact on reservoir sedimentation and resulting storage capacity losses in the Channel Reservoir and the drinking water reservoir Kaparas were tested for all management scenarios.

The scenario assessment will provide new and urgently required information on the effectiveness and applicability of the developed reservoir operation strategies.

7.2 TUYAMUYUN HYDRO COMPLEX (THC)

The Tuyamuyun Hydro Complex (THC) is located 300 km south of the Aral Sea on the territories of Turkmenistan and Uzbekistan, and is impounding the Aral Sea tributary Amu Darya. The complex (Figure 7.1) consists of four interconnected reservoirs: the Channel Reservoir (Amu Darya main stream), the Kaparas reservoir, the Sultansanjar reservoir, and the Koshbulak reservoir. Initially, THC had a total storage capacity of 7.8 km^3 but due to siltation losses, by 2001 the total storage was reduced to 6.7 km^3. Water from THC is discharged to the lower Amu Darya and to an extensive canal system supplying the regions Khorezm, Karakalpakstan and Tashauz. The reservoir complex is used to redistribute the monthly water availability and the provided storage is mainly used for agriculture (around 98%), and partly for industry and drinking water supply (up to 2%).

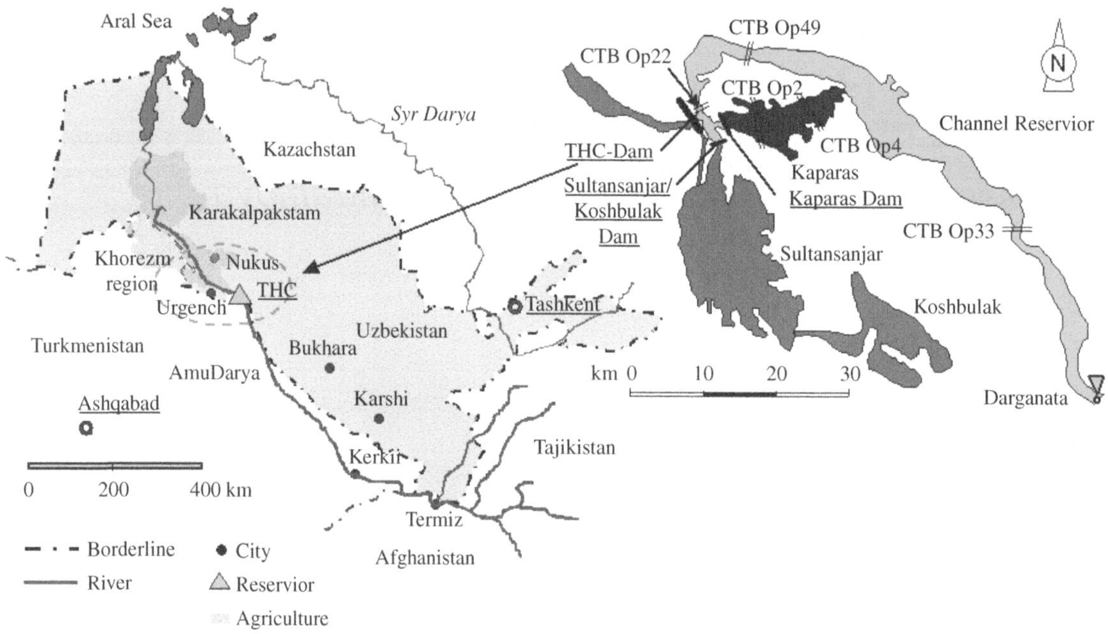

Figure 7.1. Location of the Tuyamuyun Hydro Complex (THC), with its multi-reservoir system and the inflow reference hydro post station Darganata, at the Aral Sea tributary Amu Darya.

Channel Reservoir is build by impounding the natural riverbed, whereas Kaparas, Sultansanjar and Koshbulak are designed as off stream reservoirs. From Channel Reservoir the water is either channelled into Kaparas or Sultansanjar reservoir, discharged to the downstream part of the river, or enters the different irrigation canals.

The Channel Reservoir, total design capacity of 2.3 km^3, has a design bottom level of around 110 m above sea level (a.s.l.), and a normal pool level (NPL) elevation of 130 m (a.s.l.). The operation of the Channel Reservoir is characterised by water level variations between the maximum operating level of 130 m (a.s.l) and minimum operating level of 120 m (a.s.l.). Within this range, the active regulation storage of the conservation pool is 2.1 km^3, providing seasonal streamflow regulation of the lower Amu Darya River.

Water is also abstracted from Kaparas reservoir by a pumping station, with a current capacity of five m^3/s, and is used for the centralised supply of drinking water to the regions of Khorezm and Karakalpakstan together with their main cities Urgench and Nukus.

Kaparas reservoir is connected to the Channel Reservoir near the THC dam by an open channel. Filling and release are controlled by one combined intake/outlet. Filling of the Kaparas reservoir requires a water level in the Channel Reservoir above 117 m above sea level (a.s.l.) and lower levels within Kaparas in order to initiate open channel flow. The base of the Kaparas reservoir is at 95 to 115 m (a.s.l.) and the normal pool level (NPL) is at 130 m (a.s.l.). Operation is characterized by water level variations between the NPL and the minimum operating level of 118 m (a.s.l.) with an operational storage capacity of approximately 610 million m^3, whereas the dead storage volume within the levels 95 to 117 m (a.s.l.) comprises 350 million m^3. In total, all THC reservoirs are able to provide an operational storage volume of 5.4 km^3.

7.2.1 Existing operation mode of reservoirs

Inflows to the THC multi-reservoir system are entirely dependent on the Amu Darya River discharge. The operation mode highly influences the water quantity and quality downstream of the THC. Currently, the Tuyamuyun dams are operated only for maximizing water quantity, where management rules are more based on past experience instead on formal plans. There is no formal operation plan used to manage THC reservoirs (Olsson *et al.*, 2009).

The identified characteristics of an conventional reservoir operation regime for the multi-reservoir system THC is finally summarized in Table 7.1. Existing correlations and interactions between the single reservoirs of the complex also depend largely on the overall hydrological conditions of the Amu Darya region. In short water years, both, the total water availability and therefore the operational scope, are limited.

Table 7.1. Conventional reservoir operation regimes for the Tuyamuyun Hydro Complex.

	Channel Reservoir	**Kaparas**	**Sultansanjar/ Koshbulak**
Inflow	Throughout the whole year	From September to January/February	From September to January/February
Storage	From September to January/February	March to September under dry year, variable in other years	Variable
Emptying	February to April	February to May (or shorter period in dry years)	February to June
	From May to September the flow from Amu Darya river is channelled through the reservoir without storage		
General characteristics	Provision of supply for Kaparas and Sultansanjar/Koshbulak reservoirs	Low storage volume from May to August	During wet years inflow also during summer (June, July, August)
	During wet years, possible water retention during the summer (June/July)	Filling dependant on water level at Channel Reservoir Independent release control by operational intake structure	

7.3 ADAPTATION OF RESERVOIR OPERATION FOR THE THC RESERVOIRS

Due to the expansion of the irrigated agricultural area starting in the time of the Soviet Union, the dams need to satisfy the generally increased water demand and especially the changed seasonal demand. Resulting annual variations of the stored water in THC become thereby a problem. The THC is not designed to store water for multiple year requirements, so that the operation of the whole THC multi-reservoir system under dry year conditions also has to address the monthly redistribution of water deficits.

Hence, the objectives for the adaptation of THC operation were to supply the seasonal high irrigation water demands of the lower Amu Darya region and to secure the supply of high quality drinking water by an increased use of the Kaparas reservoir, in particular for dry years.

But the main focus was on the minimization of sediment deposition in the Channel Reservoir and the maximization of the sediment transport to the lower Amu Darya stream. Furthermore, seasonal flood regulation was considered, in particular for wet years.

For determining adapted reservoir operation, the following river and reservoir characteristics were used for the development of initial and boundary conditions:

- Amu Darya flow, at inflow reference station Darganata.
  The main limiting factor for the development of new management strategies was the available inflowing water from the Amu Darya to the THC.
- Irrigation water demand of the Amu Darya Lowlands.
  Water for agricultural consumption was limited to 20.2°km^3, including 15.0°km^3 during the vegetation period and restricted by a decrease of 10–20% in dry years (regulated by ICWC).
- Drinking water demand.
  A guaranteed drinking water supply of 0,15 km^3 per year.
- Environmental flow.
  Minimum outflow to Priaralie.

- Sedimentation dynamics in the Channel Reservoir.
  A limitation of storage capacity losses for the Channel reservoir.
- Current storage capacity of the off-stream reservoirs.
  A limitation in filling and release for the off-stream reservoirs.

Due to the fact, that the THC reservoirs are interacting, the basic data for calculations were used as input for each reservoir.

7.3.1 Identification of reservoir operation based on hydrological scenarios

The development of the adapted reservoir operation based on the approach to determine optimised rule curves for different hydrological events. These events were classified in wet, median and dry year conditions.

The hydrologic regime of the Amu Darya River, with a total mean discharge of 79.3 km^3/a, can be described by an uneven temporal distribution of the annual runoff volume, with about 80% within the period April to September. High discharges occur twice a year, first for a short period in April/May after the snowmelt and second in June/July after the glacial melting in the Pamir-Alai Mountains with maximum arising.

For the historical classification of a hydrological scenario, available data related to the annual discharges within the period from 1980–2007 were summarized and categorized into extreme dry (less than 20 km^3/a), dry (from 20 km^3/a to 25 km^3/a), median (from 25 km^3/a to 45 km^3/a) and wet (more than 45 km^3/a) year flow volumes.

The lower limits of the hydrological scenario definition are shown in Figure 7.2 and identify the main proportion on mean runoff for the median year scenario for the time period 1980–2007.

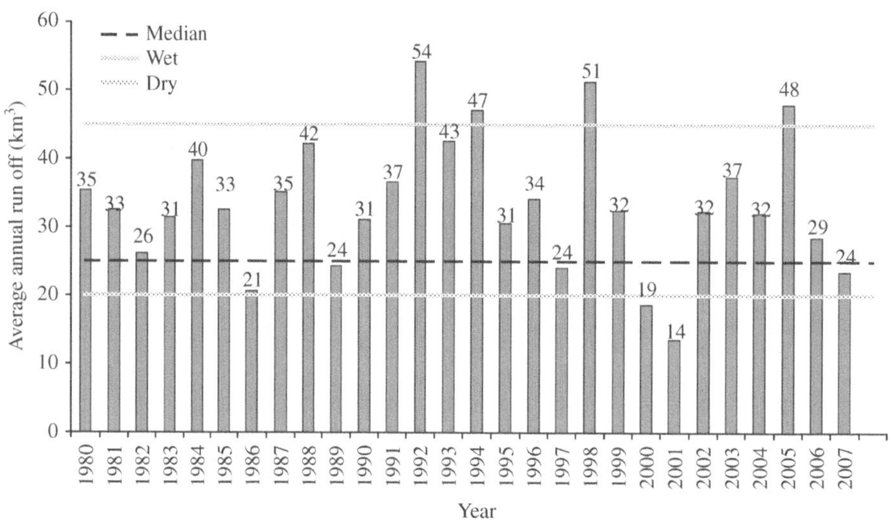

Figure 7.2. Mean annual runoff (km^3), Amu Darya River (1980–2007), Darganata station; average for dry, median and wet year.

While in wet years and median years, water quantity is sufficiently available, the water availability is limited during dry and extreme dry years, like in 2001 with 12.9 km^3.

The agricultural water supply in the Amu Darya Lowlands is regulated by the Interstate Coordination Water Commission (ICWC). The total annual consumption is limited to 20.2 km^3 in dry years, including 15.0 km^3 during the vegetation period.

During the dry years from 1980–2007 the total runoff at the lower Amu Darya River was unable to cover the annual water demand of about 20.2 km^3 (14 km^3 in 2001, Ministry of Agriculture and Water Resources of Uzbekistan). In particular the high water demands during the leaching period February to March caused an average water deficit of around 1500 million m^3 and required additional releases from reservoirs of the Tuyamuyun Hydro-Complex.

Identification of past practiced operation regimes

The measured monthly water level variations (Figure 7.3) of the THC reservoirs Channel Reservoir, Kaparas, Sultansanjar and Koshbulak in 2001, 2003 and 2005 were used for provision of past applied reservoir operation in dry, median and wet years.

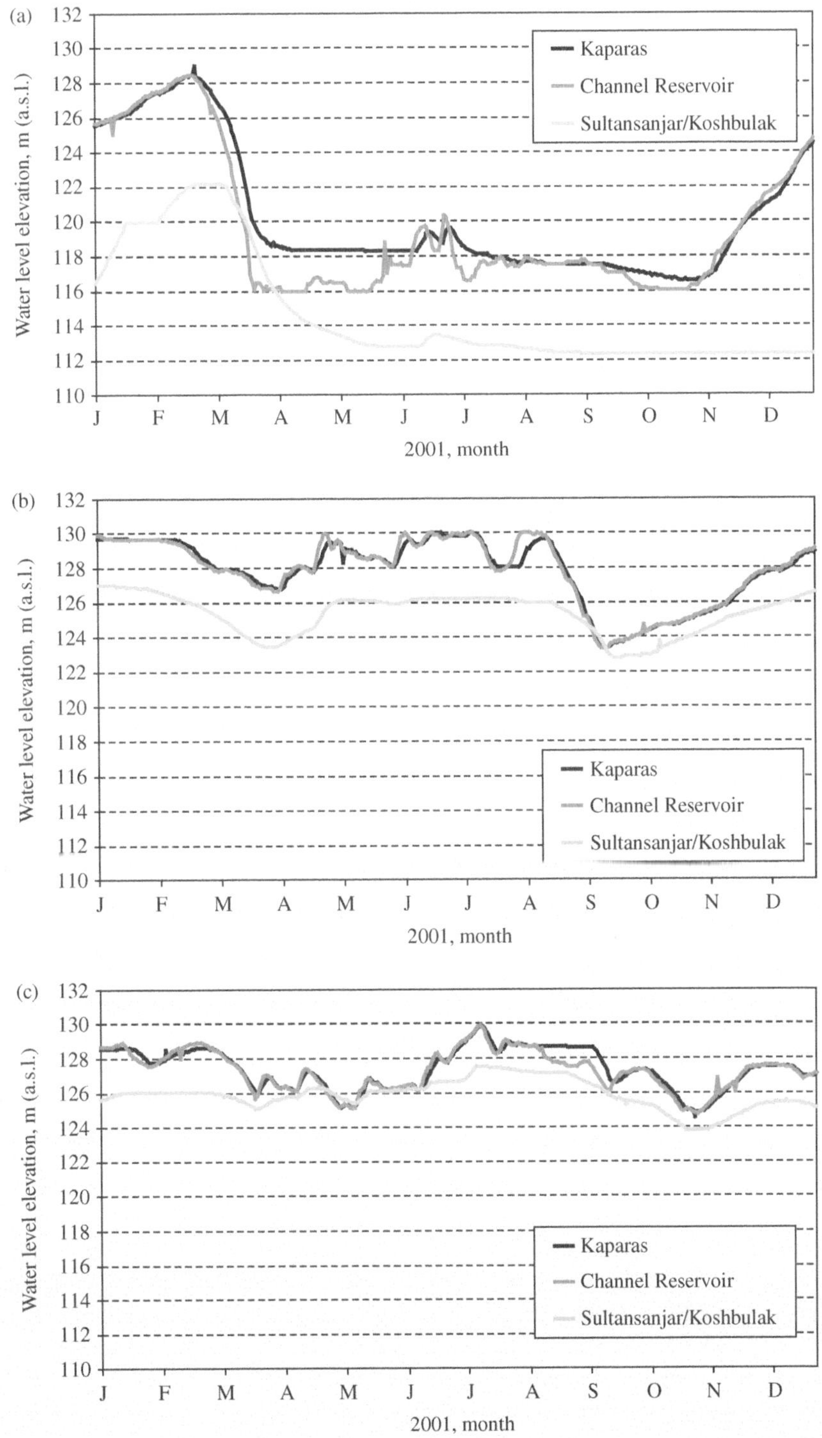

Figure 7.3. Water level elevation of the Tuyamuyun reservoirs, in the dry year 2001 (a), the median year 2003 (b) and the wet year 2005 (c).

7.3.2 Identification of reservoir management scenarios, using the *THC-model*

Improved operation strategies for all reservoirs within the THC were identified from modelling studies using the water budget THC-model. It is based on an extended water balance model, considering each single reservoir as a complete mixed reactor (Sorokin and Ikramova, 2005). The THC water balance model provides the time and volume of filling, storage, and releases of the adapted reservoir operation. The THC-model was designed to estimate the impact of changed in- and outflow for each reservoir of the Tuyamuyun Hydro-Complex (Figure 7.4) and hence providing an overall simulation for the entire complex on a monthly basis.

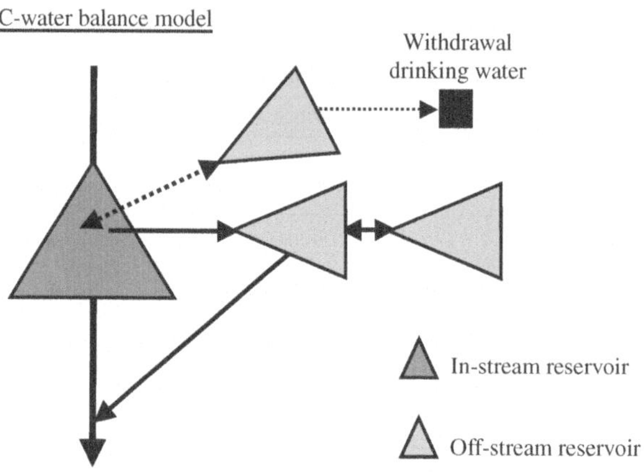

Figure 7.4. Schematic of the Tuyamuyun Hydro-Complex as basis for the THC-model.

The following terms are included in the balance calculations:

$$\Delta V_t(r) + QS_t(r) + QG_t(r) - QSO_t(r) - QGE_t(r) + B_t(r) - E_t(r) + P_t(r) = 0$$

where

∇V	:	Change in water storage volume (m^3)
QS	:	Reservoir inflow (m^3)
QG	:	Ground water inflow (m^3)
QSO	:	Reservoir release (m^3)
QGE	:	Ground-water exchange (m^3)
B	:	Run-off of reservoir catchment (m^3)
E	:	Evaporation (m^3)
P	:	Precipitation (m^3)
t	:	Time step
r	:	Modelled reservoir or lake

The daily release decision from each reservoir is based on the general rule of reservoir operation. The total operation period is T (days). ∇V is the water storage change in the reservoir r at day t and is determined by the storage volumes V from the bathymetric curve ($V_{t-1} - V_t$). $QS_t(r)$ denotes the inflow and $QSO_t(r)$ the releases to downstream or other reservoirs. Additional inflow is described by the ground water inflow $QG_t(r)$ and the additional discharge from the reservoir catchment $B_t(r)$. $QGE_t(r)$ encompasses the exchange of water between the several reservoirs by infiltration. $P_t(r)$ considers the precipitation $E_t(r)$ the additional losses by evaporation.

The obtained results from the balancing process build the basis for the further simulation process, besides which outcomes show the effects of different operation modes on the reservoir state and the downstream water supply.

Considering the objectives for the management adaptation it is planned mainly to minimize the sediment deposition in the Channel Reservoir in combination with an improved water supply for irrigation and drinking. The THC-model was used to identify the most suitable combination of water level regimes for all reservoirs of the THC for dry, median and wet year scenarios, which are indicated in Figure 7.5.

7.4 EFFECTIVENESS OF ADAPTED RESERVOIR MANAGEMENT SCENARIOS FOR MANAGING SEDIMENTS

7.4.1 Sediment dynamics modelling using the 3D MOHID-Water model

This study applied a three-dimensional reservoir model to increase understanding of reservoir sedimentation dynamics. The use of a three-dimensional model was necessary, because of the complexity of the large multi-reservoir system and the complicated sediment processes in both horizontal and vertical directions.

Sediments are a concern in many waterways and contribute mainly to losses of storage capacity. Additionally they are closely linked to water quality, because many pollutants, such as heavy metals, pesticides, and nutrients preferentially adsorb to cohesive sediments and can be transported downstream to the agricultural fields. Therefore, the study used the

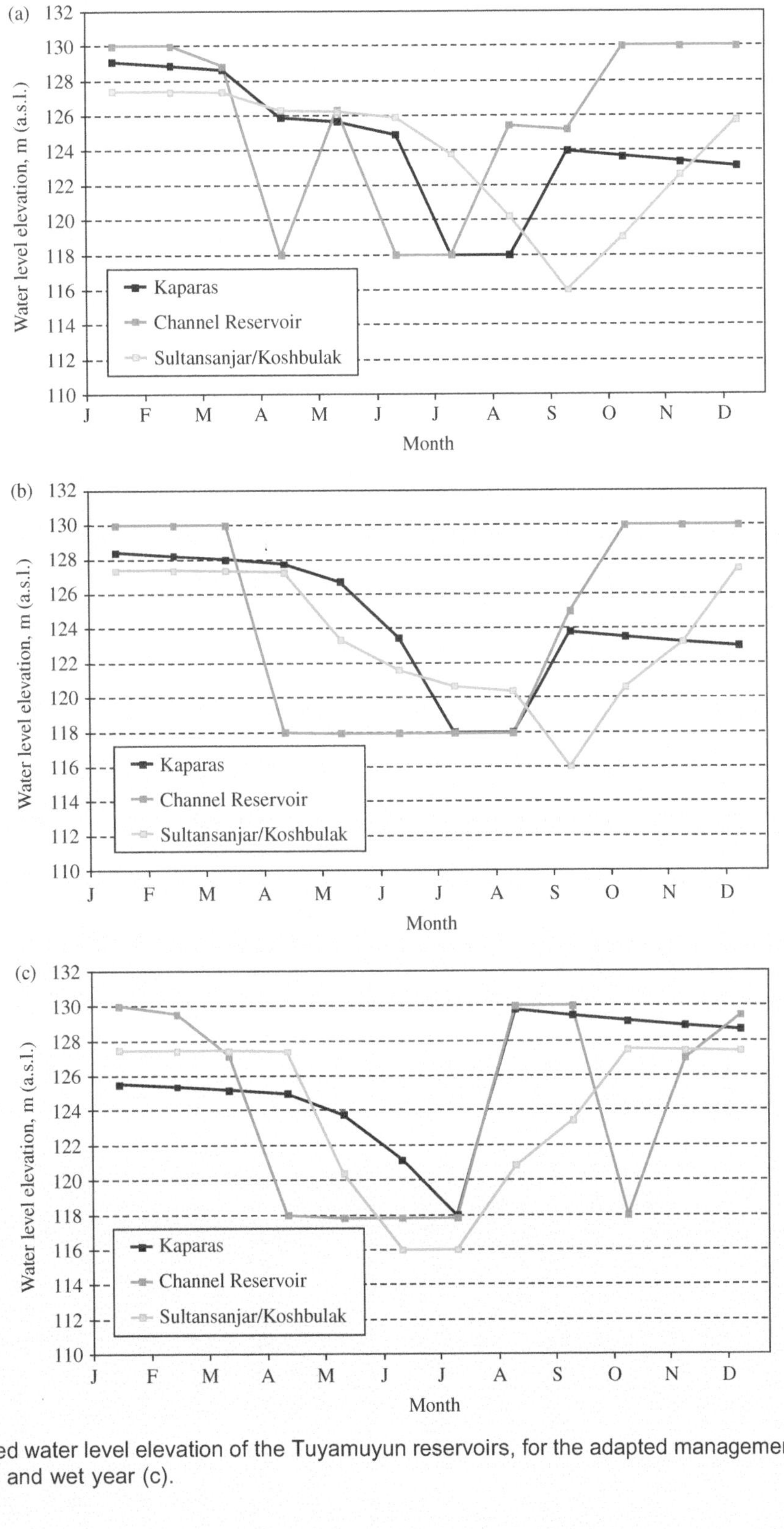

Figure 7.5. Developed water level elevation of the Tuyamuyun reservoirs, for the adapted management scenarios dry year (a), median year (b) and wet year (c).

developed three-dimensional reservoir model of the Channel Reservoir and Kaparas (Jayhun, D.5.1 and Olsson, 2009) based on the cohesive sediment transport model implemented in the water quality model 'MOHID Water'.

This three-dimensional reservoir model (Figure 7.6) was applied to increase the understanding of sedimentation processes in the Channel Reservoir, to evaluate the developed short-term management scenarios from the THC-model, and to forecast effects of storage capacity losses, which reduces the potential reserves for dry years.

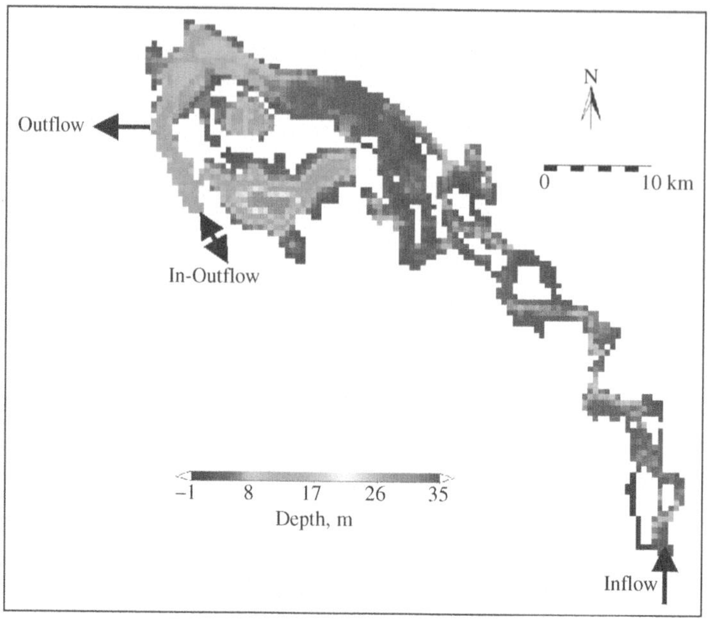

Figure 7.6. Reservoir bathymetry of the Channel Reservoir and Kaparas reservoir of the Tuyamuyun Hydro-Complex used for the cohesive sediment simulation with MOHID water.

The developed Channel Reservoir – Kaparas model was successfully calibrated and validated for hydrological characteristics and cohesive sediment dynamics in semiarid to arid conditions (Jayhun, D.5.1 and Olsson, 2009). It was applied for the period from January to December of the drought year 2001.

The model has used as boundary condition, daily discharge data from the inflow reference station Darganata for the inflow time series, daily data for the dam releases of the main dam, the Sultansanjar dam and withdrawal from Kaparas for the outflow time series. The developed time series of the cohesive sediment inflow concentration for the Channel Reservoir in dry, median and wet years based on reference information and measured sediment concentrations for the Darganata station from 1980 to 1991. Due to transition processes after the Soviet period no measured data for the sediment inflow concentration were available at the Darganata station, since 1991. Because of the lack of recent data, the effects of the operation modes have been investigated using the determined reservoir inflow and releases as boundary conditions and the established representative cohesive sediment inflow concentration (Table 7.2).

Table 7.2. Boundary conditions for cohesive sediment simulations, developed average cohesive sediment inflow, 2001/dry year, 2003/median year, 2005/wet year.

	Monthly sediment concentration, mg/l 2001/dry year	**2003/median year**	**2005/wet year**
January	834	925	960
February	817	561	1453
March	516	994	1564
April	467	2379	1789
May	606	2967	4420
June	1640	4706	4953
July	910	3211	4850
August	665	2037	3099
September	490	965	1772
October	302	622	1051
November	344	682	937
December	495	928	1057

The basic settings determining initial conditions for the simulation using the MOHID water model are summarized in Table 3. The application used as basic input data the measured reservoir bathymetry from 2005 and measured sediment concentrations of 6 mg/l, recorded at the middle section of Channel Reservoir in January 2001. The simulation period for all scenarios was from January 1st to December 31st. The simulation of the reservoir operation started with the initial water level in the Channel Reservoir nearby the main dam. Past records of climate data for temperature, relative humidity, cloudiness and wind from 2001, 2003 and 2005 were used as input for the model estimation of the potential evaporation.

Table 7.3. Basic settings used for cohesive sediment simulations of Channel Reservoir – Kaparas in dry, median and wet years.

	Channel Kaparas Reservoir system					
Conditions	**Dry**		**Median**		**Wet**	
Conventional/ adapted operation	2001	Dry scenario	2003	Median scenario	2005	Wet scenario
Start water level m (a.s.l.)	125.60	130.00	130.00	130.00	128.68	130.00
Simulations period	1 Jan. 2001–31 Dec. 2001		1 Jan. 2003–31 Dec. 2003		1 Jan. 2005–31 Dec. 2005	
Weather/climate data	2001, Urgench		2003, Urgench		2005, Urgench	
Reservoir bathymetry	Measurements 2005					
Initial sediment concentration	6 mg/l, measurements from January 2001					

7.4.2 Simulation results for 2001 and dry year scenarios

The simulation results for the sediment deposition in the Channel Kaparas Reservoir system are depicted in Figure 7.7 conventional (2001) and adapted (dry year) operations scenarios. Both graphs for the conventional (black line) and the adapted (grey line) operation show a similar tendency and characteristic. There is a continuous increase up to 500,000 t from January to the first part of March, followed by a period with a lower gradient up to 640,000 t until June. In the second part of the year from June to September, both graphs have a high increase up to 1,900,000 t. Followed by a different decline of the graphs gradient, the results of the total annual sediment deposition show 2,224,866 t for the black line and 2,317,504 t for the grey line at the end of December.

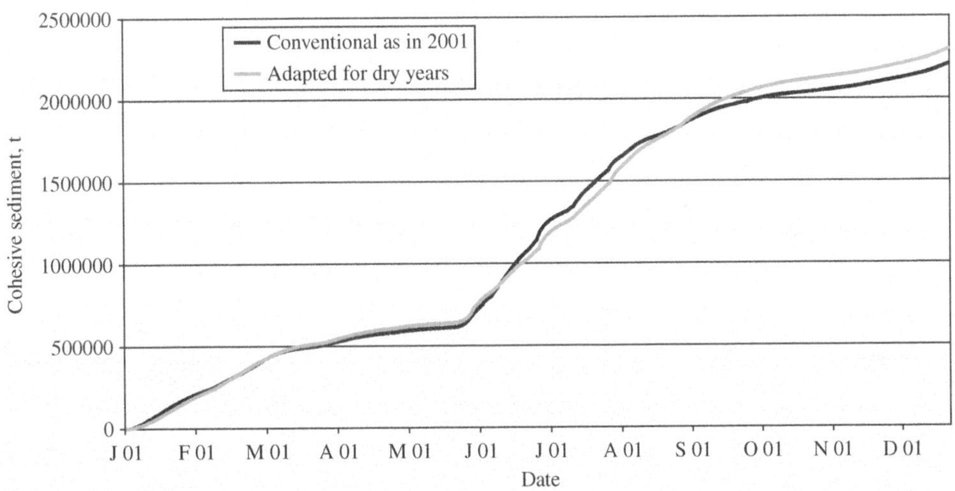

Figure 7.7. Simulation results for the sediment deposition in the Channel Kaparas Reservoir system for the scenarios of conventional operation as 2001 and the adapted operation for dry years.

The total annual sediment deposition in the Channel Kaparas Reservoir system for the simulated dry year operation has an increase of 92,638 t (4%) compared to the conventional operation in 2001.

7.4.3 Simulation results for 2003 and median year scenarios

The simulation results for the sediment deposition in the Channel Kaparas Reservoir for the 2003 and median year management scenarios are shown in Figure 7.8. From the beginning of the year until April, they have the same tendency with similar values. The graph of the conventional 2003 scenario (black line) describes a continuous decrease to 1,200,000 t from January to the middle of April. In the second third (mid of April to September) there is a rise with a higher gradient up to 14,400,000 t. The last third presents a decline of the gradient, which ends in a total annual deposition of sediment of 16,064,988 t.

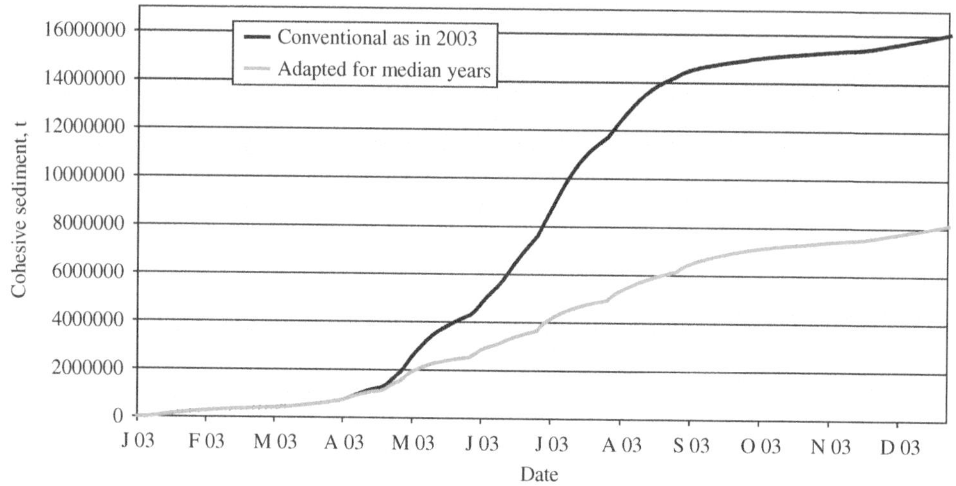

Figure 7.8. Simulation results for the sediment deposition in the Channel Kaparas Reservoir system for the scenarios of conventional operation as 2003 and the adapted operation for median years.

The graph (grey line) representing the result for the adapted scenario median years, describes a continuous decrease up to 1,200,000 t from January to the mid of April. Compared with the conventional graph followed a rise with a low gradient up to 7,200,000 t till October. The final part to December has a lower gradient which results in a total annual sediment deposition of 8,132,965 t.

The results indicate that there is a decrease of 51% (minus 7,932,023 t) of the annual sedimentation for the simulated adapted operation, compared to the simulated conventional operation.

7.4.4 Simulation results for 2005 and wet year scenarios

Figure 7.9 shows the simulation results for the sedimentation in the Channel Kaparas Reservoir of the conventional (black line) and adapted (grey line) operation scenarios. They have the same incline from the beginning until April followed by a parallel gradient with same tendency. In the months from January to mid June the black graph increases up to approximately 5,400,000 t and the grey graph rises up to 4,500,000 t for the sediment deposition. This part is followed by a period with a higher gradient until September (sediment deposition for the conventional operation with 14,000,000 t and for the adapted with 12,500,000 t). From September until the end of the year the increase is with a lower gradient. The conventional graph ends with a total annual deposition of sediment of 16,202,056 t and the grey for the adapted operation with 14,799,354 t.

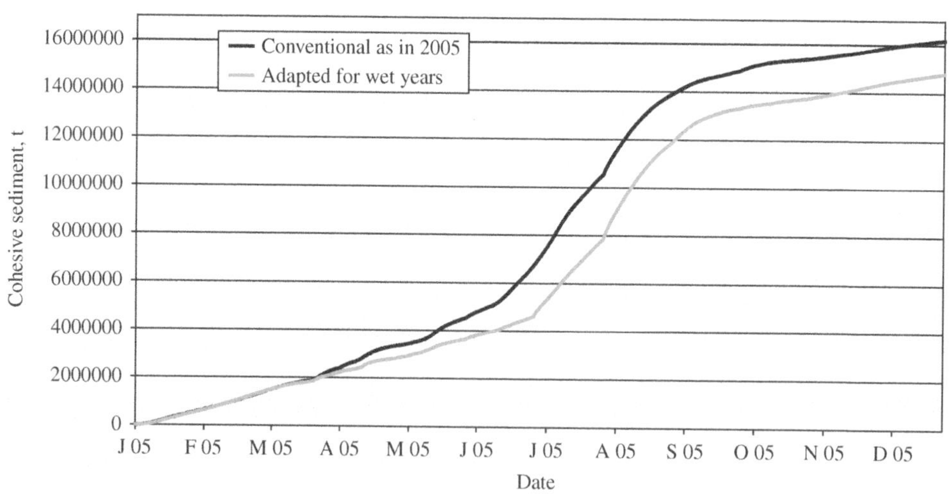

Figure 7.9. Simulation results for the sediment deposition in the Channel Kaparas Reservoir system for the scenarios of conventional operation as 2005 and the adapted operation for wet years.

The total annual deposition of sediment in the Channel Kaparas system for the simulated adapted operation has a minor decrease of 1,402,702 t (8.5%) compared to the conventional operation with 16,202,056 t.

7.5 CONCLUSIONS

This study assesses the impact of the proposed management strategies on the sediment dynamics and resulting storage capacity losses. The assessment results show that the Channel Reservoir, significantly influence the sediment regime of the lower Amu Darya by acting as a sediment trap. The main risk factor for the Channel Reservoir storage capacity is the high inflow sediment concentration (up to 4500 mg/l) in high water years, which describes a three times higher sediment concentration as during the low flows in dry years.

This is also demonstrated by the simulation results for the Channel Reservoir sediment deposition for all management scenarios under dry, median and wet conditions. It is shown that the dry year operation has the smallest impact and the wet year operation has (with four times higher annual sedimentation as in dry years) the highest impact on annual reservoir storage capacity losses.

For the particular system Tuyamuyun Hydro-Complex studied, it is shown that adapted reservoir operation scenarios, developed to improve water quantity and quality conditions for downstream use, has compared to the present practise a significant effect on the sediment deposition in the Channel Reservoir. The simulation results identified for the adapted management scenarios a reduced sediment deposition of 51% under median conditions, of 15% under wet conditions and no significant improvement under dry conditions.

As the characterisation of hydrological scenarios identifies that the frequency of median year was higher than wet or dry years during the investigated time period, the high reduction of sediment deposition under median conditions results in an important contribution for securing reservoir storage capacities.

It is expected that due to climate change impacts the frequency of dry and wet years will increase. Therefore the impact of these extreme scenarios is more and more important for the daily reservoir operation. Nevertheless, due to the low sediment deposition rates under dry year conditions the dry year management scenario has a minor impact on the total reservoir storage capacity losses. But this study also shows with the adapted reservoir operation for wet years a significant improvement.

7.6 REFERENCES

Basson, G.R. and Rooseboom, A. (1997) Dealing with Reservoir Sedimentation. Water Research Commission Report No. TT 91/97.

Jayhun, Deliverable report 5.1 (2009) IMPLEMENTATION OF MODELLING TOOLS. JAYHUN Interstate Water Resources Risk Management. “Towards a Sustainable Future for the Aral Basin,” Specific Targeted Research Project Contract No. 516761.

Morris, G. and Fan, J. (1998) Reservoir Sedimentation Handbook. McGraw-Hill, New York.

Olsson, O. (2009) Combined Operation and System Assessment of Reservoirs for management under water stressed conditions during dry years. Publications of the Institute of Water Quality and Waste Management. (Veröffentlichungen des Institutes für Siedlungswasserwirtschaft und Abfalltechnik der Universität Hannover.)

Olsson, O., Ikramova, M., Bauer, M. and Froebrich, J. (2009) Applicability of Adapted Reservoir Operation for Water Stress Mitigation Under Dry Year Conditions. Water Resources Management, DOI 10.1007/s11269-009-9446-x (available online).

Sorokin, A. and Ikramova, M. (2005) Modelling studies Tuyamuyun Hydro Complex model. IWMT ‘Development of integrated water management tools for the Tuyamuyun reservoir complex – Improvement of drinking water supply and public health in the disaster zone of lower Amu Darya’, Final Report EU-INTAS 1043/DFG 1607/FR1-1.

Chapter 8

Natural and manmade risks to water resource security in the Amu Darya Basin

Trevor Tanton, Malika Ikramova and David Rycroft

8.1 FACTORS AFFECTING THE RISK POTENTIAL

We know what the annual water resources of the Amu Darya basin have been, the mean discharge being approximately 66 km^3 $year^{-1}$ (based on data 1980–2005), but we do not know what they will be in the future. The records show that as well as random variation between years there are also sequences of years with above and below average flows, with flow falling to 46 km^3 in the driest years and rising to 109 km^3 in the wettest. Most of the mean annual flow is abstracted and hence in the drier years the irrigated area has to decline whilst in addition many irrigated crops are grown under water deficit conditions. As an adaptation measure the river basin authorities implement agreed water management procedures to try and share the economic loss throughout the region although in reality, and as is always the case, the crop losses are highest for the tail end farmers, be they at the lower end of the river basin or at the end of the distribution canals. In a state run system the farmers whose irrigation makes a loss because of centrally planned deficits may not suffer financially if water is distributed for the overall good of the State, since farmers experiencing water shortages can be compensated at the expense of those who receive water; but if farms are privatised, then farmers who do not have water will either cease farming or live with an unreliable source of income, leading to poverty, starvation or land abandonment. It can clearly be seen from this example that management of risk associated with natural variability in water supply is not just an engineering problem but is linked to the very basic economic system under which irrigated farming operates. This single example is used to demonstrate the multi-faceted nature of water risk and the complexity of the problem that needs to be understood and addressed if we are to provide robust water resource management.

8.1.1 Risk and water resource management

A number of factors can affect the future supply of water in a river: for example if one out of three years a flow of 50 m^3 s^{-1} is exceeded there is a risk that in 2 out of 3 years there will be less than 50 m^3 s^{-1}. With adequate flow data this sort of risk can be quantitatively assessed in both the short and medium term using well-understood stochastic principles. There are, however, also unquantifiable risks to water resources such as events that occur far beyond the historic data sets on which risk can be reliably quantified. These may be divided into two types of risk: those arising as a result of 'acts of God' and those resulting from the 'acts of man'. Acts of God include such events as extreme flood or drought. Such risks can be catastrophic in valleys whose rivers arise in high mountains, particularly those associated with high levels of seismic activity. Examples of such events can result from surging glaciers, earthquake triggered land/mud slides, dam failures, and from heavy snows followed by rapid thawing associated with high levels of rainfall. Risks resulting from the 'acts of man'

 ISBN: 9781843393085. Published by IWA Publishing, London, UK.

can conceptually be managed, but there are many examples of large irrigation systems that have failed in the past because of human activities. This may be because of war, economic collapse, the taking of the water by upstream riparian inhabitants, the failure of good governance at the national and/or project level, salinisation, and to a lesser extent dam failure. Many of these issues relate directly to the large irrigation schemes in the Amu Darya basin, where economic collapse in the 1990s and institutional decline has resulted in poor water management and both inadequate maintenance and investment in infrastructure. As a result, and because of a severe shortage of effective subsurface drainage systems, between 1993 and 2003 an estimated 600,000 ha of irrigated cropped land in Central Asia had became derelict as a result of salinisation, whilst Uzbekistan alone loses 20,000 ha annually as a result of salinisation.

Climate change has recently introduced another man-made risk affecting the reliability of water resources. Unfortunately the magnitude and the number of poorly quantifiable risks are greater than the quantifiable ones making it impossible to stochastically forecast quantifiable risk in a reliable manner.

Although we know many of these risks exist, and that they can be catastrophic, their frequency cannot be reliably predicted and because they may occur very infrequently or at an indeterminate time in the future, they are pushed to the back of the mind. With the exception perhaps of salinity, it is very difficult to reliably identify the frequency or magnitude of such events; but this does not mean that it is not possible to take steps to try to manage them, even if they not precisely quantifiable. Provided we have a concept of the likely risks, their causes, likely frequency and magnitude it is possible to take them into consideration in developing robust sustainable water resource utilisation plans. The planning approach needs to ensure that water resource infrastructure is able to accommodate all the major risks that are considered likely to occur and to ensure that they are manageable.

This study evaluates a number of these risk factors in the context of the water resources of the Amu Darya Basin and considers the measures that might be taken to mitigate them. The future risk to the water resources of the Amu Darya basin has been assessed for the year 2085. Although this may appear to be far in the future, and critics might say we should have considered a shorter time frame, it was chosen as a time that many of the children born this year (2009) will witness in their lifetime, while it is also a standard period used in the IPPC's assessment of climate change.

8.1.2 Risk posed by climate change

It is widely accepted that climate change is happening and that the most likely cause is the anthropogenic release of greenhouse gasses. To facilitate investigation of the potential impact of climate change a range of greenhouse emission scenarios that could occur in this century has been internationally agreed. Climate modelling groups across the globe use them in global climate circulation models that they have developed to simulate temperature and precipitation outcomes for the different emission scenarios up to 2085. Each country tends to favour its own model: for example the United Kingdom favours the British Meteorological Service model HadCM3 whilst Australia favours its CSIRO model. The models are all based on a computational cell size typically of about 3° to 5° latitude and longitude across the earth's surface.

Methodologies exist for 'downscaling' the models outputs to smaller cells, in some cases to an area of 250 km^2 or less, by including such factors as topography and historic data sets in order to improve the interpolation of temperatures/precipitation for each location. The output is usually in the form of seductively coloured maps of potential future climatology up to 2085. The idea is that these maps can then be used for planning and design of adaptation measures for climate change. The problem is that downscaling implies improved accuracy, but this could not be further from the truth. All it provides is an apparent increase in precision. This is because of the inherent sources of error in the original models and scenario assumptions that arise from our incomplete understanding of atmospheric physics, inherent randomness within the climate systems, initial starting conditions, computational shortfalls, and the inability to predict mankind's economic development and impact on the world. It is therefore not surprising that there is considerable variation between the different model outcomes at both the world and local scales, even for the same climate scenarios: see for example the surface temperature anomalies, climatologies, and monthly temperatures respectively derived from 6 Global Circulation model outcomes for the A2 climate change scenario in the IPCC 4th Assessment for Bukhara in 2085 as shown in Figures 8.1, 8.2 and 8.3.

It is important to note in Figure 8.3 that despite the 2°C (2.7°C in summer) differences between the outcomes of the six models for the A2 scenario there is appreciable agreement between them, with all models showing a rise in temperature for Bukhara above the reference year of 1975 of about 5 to 6°C for June to August 2085. Unfortunately, the variability in predictions does not stop at the differences between the global circulation models but also extends to the emission scenarios. If the models are applied to the B2 climate change carbon dioxide emission scenario they typically indicate a mean temperature rise of 4 to 5°C for the same period. The A2 and B2 scenarios represent a fairly wide range of potential scenarios and range of future greenhouse gas levels. It is therefore concluded that the temperatures across much of Central Asia in summer in 2085 are likely to be between 4 and 6°C hotter than the reference temperature of 1975.

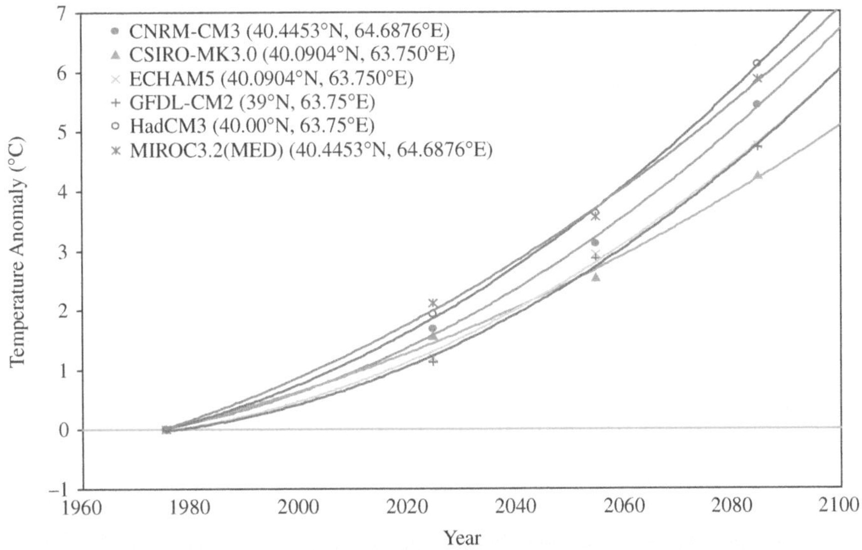

Figure 8.1. AR4 SRA2 scenario, 30 year mean surface temperature anomalies (Jun–Aug Average, 2010–2099) relative to the Climate Research Unit of the University of East Anglia, UK (CRU) observed data for the period 1961–1990 for Bukhara (39.75°N, 64.25°E). MODELS INCLUDE: CSIRO model, Australia's Commonwealth Scientific and Industrial Research Organisation; ECHAM5 model, Max Planck Institute for Meteorology; GFDL-CM2 model, Geophysical Fluid Dynamics Laboratory model; NOAA model, USA; HAD CM3 model, the British Meteorological Service Hadley; MIROC 3.2 (MED) model, Centre for Climate System Research (CCSR), the University of Tokyo.

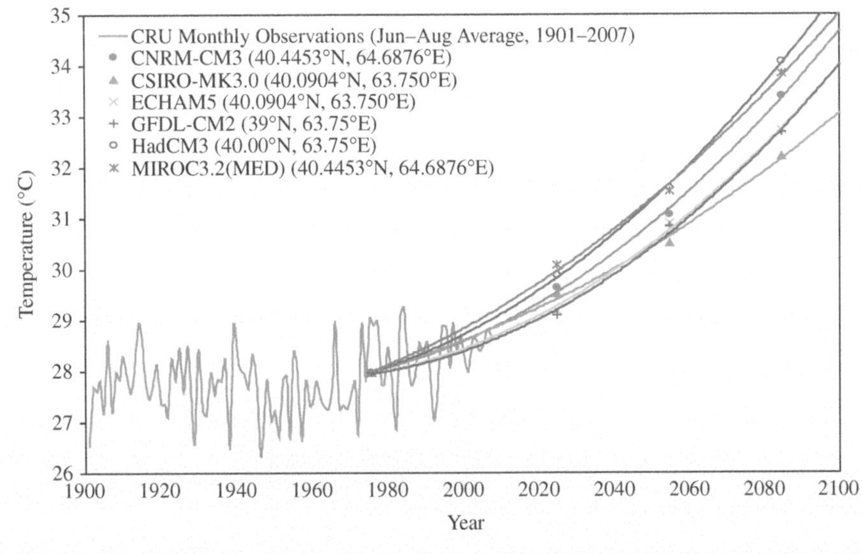

Figure 8.2. AR4 SRA2 outcomes, 30 yr 2 m mean surface temperature climatologies (Jun–Aug Average, 2010–2099) for 6 climate change models based on CRU (1961–1990) observed data for Bukhara (39.75°N, 64.25°E).

It is important to note that all Global Climate Circulation Models and climate change scenarios indicate that Central Asia will become appreciably hotter than at present, and that the main body of models tend to show some agreement for a given scenario, being about +5°C±1°C for Bukhara by 2085. At high altitude, as the thickness of the air insulation blanket thins and its effectiveness diminishes, the increase in temperature will be much less and is expected to be only 1 or 2°C warmer. Savitskiy *et al.* (2008) have analysed the increase in mean observed daily temperature from six climate stations in the region and they have established that between 1975 and 2005 the mean daily temperature in the region has already increased by about 1°C. This recent increase in temperature in the middle and lower basin indicates that an increase of 3 to 5°C by 2085 is highly likely if the present trend continues.

All the climate change models and scenarios predict little change in precipitation in the basin, e.g. 0±5%, but this estimate is associated with a high degree of uncertainty. This can be seen in Figure 8.4 which depicts the monthly

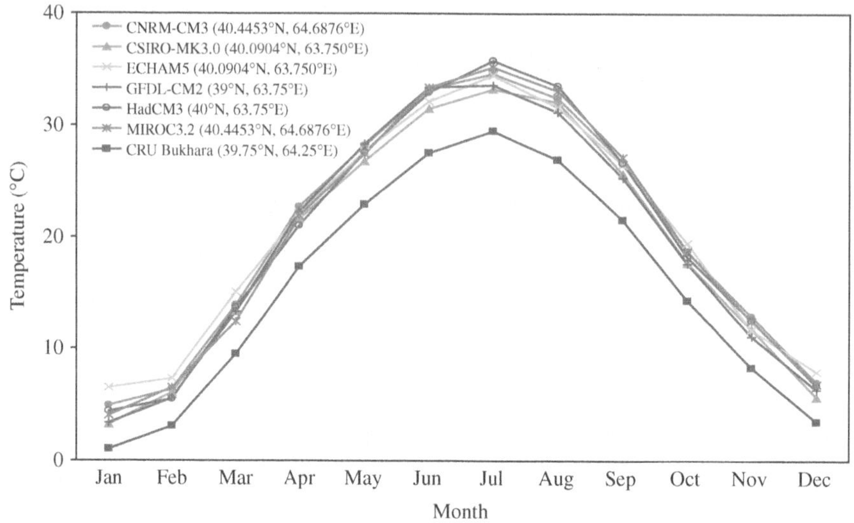

Figure 8.3. AR4 SRA2 scenario, 30 yr 2 m mean surface temperature monthly climatologies (Jan–Dec, 2070–2099) based on 6 models and CRU (1961–1990) observed data for Bukhara (39.75°N, 64.25°E).

precipitation anomalies for 7 Global Circulation Models (GCMs) for Dushanbe (capital of Tajikistan). This variability is not surprising since GCMs are at their weakest in mountainous regions where local advective conditions play such an important role in influencing precipitation. Hence, GCM forecasts of future precipitation in mountainous regions must be considered with great care.

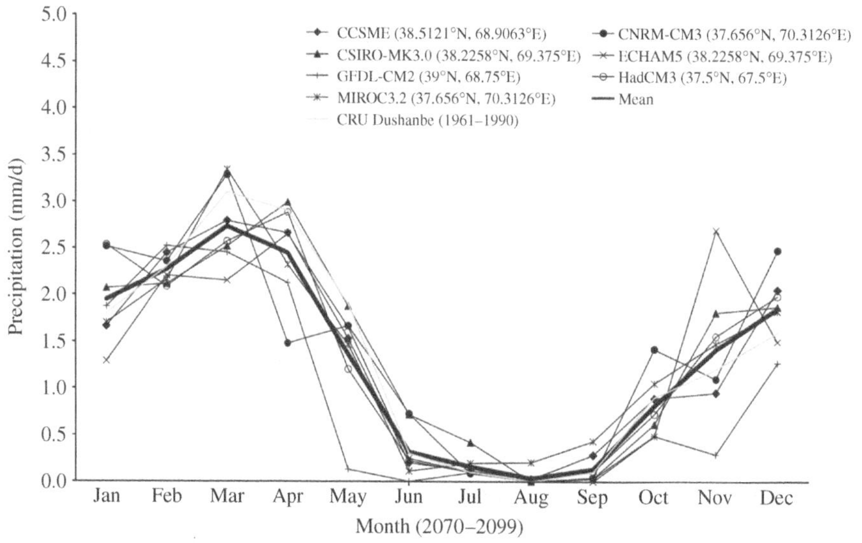

Figure 8.4. AR4 SRA2 30-year precipitation climatologies (Jan–Dec, 2070–2099) downscaled using the 1961–1990 baseline normal for Dushanbe (38.25°N, 69.25°E).

A more complete description of future likely climate change in the Amu Darya basin resulting from the IPCC's 4th assessment is provided by White and Tanton (2008; JAYHUN report to D.3.2 and Chapter 2).

The models when considered as a group provide some indication of the likely future climate and an indication of the range of uncertainty. Although this does not provide a measure of stochastic certainty it does allow the water resource planner to develop robust water resource plans for the future. Managing the risks to the water resources associated with climate change cannot be considered in isolation, however, but needs to be integrated into an overall assessment of total risk from all sources. Planning tools can then be devised that allow the development of a long term rational and sustainable water resource management strategy that is sufficiently robust to accommodate all risk factors.

The question then arises, if we have appreciable uncertainty concerning the level of climate change and that uncertainty becomes greater the further we look into the future, how do we adapt if we are to minimise the adverse impacts on both society and the environment? The approach is to be found in earthquake engineering; we do not know when the next earthquake will occur or even if it will occur, nor how severe it will be. The approach adopted to manage earthquake risk in most countries is to identify from historic seismic data the locations where earthquakes are likely to occur and to then make an informed decision of the likely maximum size. The engineers then apply their skills to minimise the potential adverse effects. Politicians then have to make a decision, hopefully based on the best advice available, on what measures can be taken to try to manage the unquantifiable risk, given the economic circumstances of the region. In a wealthy area, such as California, the local government can require and achieve very high construction standards to ensure earthquake robustness of infrastructure in order to minimise risk; but in areas like Afghanistan or Northern Pakistan, because of their weaker economies, it is not within the government's power to impose such high construction standards, and hence the risk of catastrophe is much greater.

Engineers can develop similar robust water resource management approaches that guarantee a given level of security for the water resources of an irrigated basin; unfortunately political pressure is often applied to increase the irrigated area to meet an ever-growing demand for irrigated land, thus negating the plans and reducing the overall level of security for the existing users in the basin.

8.2 CLIMATE CHANGE AND ITS POTENTIAL EFFECT ON THE WATER RESOURCES OF THE AMU DARYA BASIN

8.2.1 Temperature and future potential evapotranspiration in the cropping season

Bukhara and the surrounding irrigation massifs lie at the centre of the Amu Darya Basin and have a climate that is fairly representative of most of the irrigated parts of the basin. The downscaled temperature anomalies for the A2 scenario derived from the IPCC's 4th assessment (IPPC 4th assessment 2008) for the HadCM3, CNRM-CM3, CSIRO-MK3.0, ECHAM5, GFDL-CM2, and MICROC global circulation climate models for Bukhara are shown in Figures 8.1, 8.2 and 8.3.

It can be noted that the observed increase in temperature in Bukhara since the reference date of 1975 is consistent with the mean prediction of these models. Although there is a high level of uncertainty looking so far into the future, the models (Figure 8.2) and main IPPC future scenarios do suggest that by 2085 we can expect a mean summer temperature increase in the growing season in Bukhara and the other main irrigated areas of the Amu Darya basin of $+5°C \pm 1.5°C$, that spring will start a month earlier and autumn end about 2 weeks later. This results in a calculated increase in the crop water demand (potential evapotranspiration) for the region of +9 to +19%, with a mean of 14%. This equates to a mean additional demand for water from the reference year (1975) of between 5 and 12 km^3 $year^{-1}$ for the basin's entire irrigated area. By 2085 to irrigate the existing area at the same level of efficiency will require an additional 5 to 12 km^3 $year^{-1}$ of water: since nearly all the river water is currently abstracted for irrigation, this is clearly not available. Hence the irrigated area must contract by on average 14% or some 250,000 hectares simply to maintain the *status quo* for reliability and to prevent further uneconomic yields from occurring on the irrigated land.

The A2 emissions scenario used in the analysis is just one of a number of scenarios for Carbon Dioxide emission outcomes. The authors consider that it is probably the most realistic, however, as it is the one that most closely matches both human behaviour and past precedents when it comes to shared common resources. Nevertheless it is only one possible scenario, but it is important to note that all future scenarios used by the IPCC predict significant increases in temperature in the Aral basin that are well above the anticipated global average temperature increases, hence all result in a very significant increase in crop water requirements by 2085.

With global warming the potential cropping season in the Amu Darya basin will be longer, with the planting season starting some 4 weeks earlier in spring and the growing season ending two weeks later in autumn. This is expected to result in increased cotton yields due to the longer growing season but also increased demand for water. On the other hand the shorter winter season and the hotter spring and early summer temperatures could restrict grain filling of winter wheat due to its high temperature sensitivity. Hence, although there are threats there are also some opportunities as a result of climate change.

8.2.2 Effect of climate change on the availability of surface water resources

All the global circulation models in IPCC's 4th Climate Change Assessment Report anticipate little or no change in precipitation in the Amu Darya basin by 2085, all indicating less than a ±5% change from the existing levels of precipitation. Away from the mountains very little precipitation falls on the irrigated massifs located on the plains and hence only a relatively small percentage can be considered to be effective in contributing to crop water requirements. As a result, the predicted change in precipitation equates to an insignificant change in the demand for water for irrigation

The sources of the region's waters are the high mountains of the Pamirs in Tajikistan and the north western slopes of the Hindu-Kush in Afghanistan, with over 800 mm of annual precipitation falling as snow at the higher altitudes. This compares to the arid steppe where up to 300 mm $year^{-1}$ falls close to the foothills, but less than 100 mm $year^{-1}$ out on the open steppe. The global circulation models all indicate that by 2085 there is likely to be little change in precipitation throughout the basin, although the mean of all models does indicate a small downward trend. Local advective conditions are of major importance in determining precipitation in mountainous regions, and Climate Global Circulation Models are at their weakest predicting precipitation outcomes in such regions, hence any model forecasts must be considered with great care. The snow and glacial melt from the Pamir and Hindu-Kush mountains contribute about 85% to the region's water supply, but due to a lack of sound hydrological data the hydrology of these mountainous areas is still poorly understood.

Although annual precipitation is not predicted to change significantly (the mean of 7 GCM outputs for the A2 scenario for 2085 indicates a decrease of about 5%), the volume of runoff in the mountains and its temporal distribution could change quite appreciably. By 2085, the expected increase in mean temperatures at lower altitudes will be around $+5 \pm 2^\circ$C, although at higher altitudes only +1 or 2°C. The winter will start some 2 weeks later and the thaw some 4 weeks earlier, so appreciably more of the annual precipitation is expected to fall as rain rather than as snow, allowing more to infiltrate into the soils. The vegetation line will migrate upwards with the ablation zone by an estimated 500 m. This will result in higher losses of water due to increased evaporation and sublimation, and hence will result in less of the precipitation being available to form runoff. The magnitude of the associated reduction in river discharge is very difficult to quantify. The existing empirical runoff models used to predict runoff in the upper basin are all based on reliable stochastic methodologies and are soundly calibrated, but these future changes are likely to invalidate them due to the changing physical nature of the catchment.

In common with all glaciers in the world, the glaciers of the Pamirs and Hindu-Kush are shrinking and it is generally considered that the rate of loss in recent years has been accelerating. Although a considerable amount of work has been published on this topic, a lack of sound data limits our understanding and it will be many years yet before we fully understand the complexity of the system. It is, however, possible to make some general conclusions. In the Pamirs several estimates have been made of the rates of glacial loss. Konovalov and Schetinnicov (1980) estimated that in 1980 the glaciers contained some 468 km^3 of water and that the loss of glacial volume during the previous 23 years had been 56 km^3 or 0.5% per year (2.7 km^3 per year), although this estimate has a high level of uncertainty. Bedford (1997) reviewed the work of 4 researchers who used different approaches to identify the rates of glacial loss in the Pamirs. The mean of the four approaches indicated that the annual loss was about 2.2 km^3 or 0.4% per year. It would therefore indicate that at the present time unsustainable glacial loss (permanent loss due to glacial retreat) only contributes about 3% to the annual discharge. Although in years of higher than average precipitation glacial melt is not of great significance, in dry years it is much more important since in these years unsustainable glacial melt can contributes up to 5% of the total river discharge. Hence, if unsustainable glacial melt (glacial retreat) were not taking place the mean annual flow would only decline by 3%, provided that the other hydrological aspects of the basin remain the same.

Sustainable glacial discharge and high altitude summer snow melt are much more important in contributing to the annual flow. Konovalov (2007) indicated that glacier and high altitude snow melt contributed 26 and 21% to their respective mean annual flows. Provided precipitation remained the same, significant loss of the glaciers may not result in a large difference to the mean annual river discharge but it will result in much greater variation in annual discharges, and as a result, in extreme dry years greater water shortages are to be expected throughout the basin. Konovalov showed that in dry years during the summer irrigation season, glacial melt can contribute up to 57% of the flow of the Panj and 39% of that in the Vaksh Rivers, and hence the sustainable glacial melt plays a very important role at times of water scarcity and is essential in buffering the river discharge in years of low precipitation. For example, in the most recent driest year, the total annual river discharge was 46 km^3 compared to the average of 66 km^3. Discharge emanating from glaciers will not completely disappear as a result of global warming, but with the current rate of loss of 0.4% per year, by 2085 it is conceivable that glacial discharge could fall by about 30%. This would mean that in an equally dry year by 2085 discharge could fall to around 30 km^3y^{-1}. Such a level of discharge will be catastrophic for the farmers of both Uzbekistan and Turkmenistan.

We therefore find ourselves looking at a time of great uncertainty as to the future availability of the water resources of this once great river and its ability to meet the needs of the populations of the riparian countries. Over time our insight into what is happening will become clearer, but if we wait for a comprehensive understanding of the issues before we act it is likely to be much too late. From the above it can be concluded that there is likely to be appreciably less water available in the future and we have some idea of the potential magnitude of the risk, i.e. by 2085 we could see a reduction in the mean average water supply of 5 to 15%, but in the driest years flow this could reduce by up to 35%. Although statistical certainty cannot be placed on these estimates the evidence and logic dictates that the threat is very real. It would be foolhardy indeed to ignore this risk in the future planning of the basin's water resources if they are to equitably meet the future needs of the inhabitants of the basin. It is, however, possible to develop plans to manage such risks without knowing their exact severity. The risk needs to be integrated into a comprehensive adaptation/risk management strategy for the water resource development of the basin as a whole, since in the worse case by 2085 with the combination of increased evaporative demand and decline in supply there is a possibility that in extremely dry years water supply for irrigation could only meet 50% of demand.

8.2.3 The future of irrigation in the Amu Darya Basin

The current demand for water in the Amu Darya basin includes water needed to irrigate the estimated 3.4 million ha of crops and water needed by humans and industry. River basin modelling studies using the water management modelling software WEAP [AMU-WEAP-100 (see Chapter 5)] indicated that the mean theoretical annual crop water demand for the current irrigation command area of the Amu Darya Basin (in 2008) was 34.5 km^3 and an additional 4.8 km^3 was required for municipal and industrial water supply, yielding a total annual demand of 39.3 km^3. This estimate was based on the IPCC mean climate data for the reference period 1961–1990 (1975). The annual mean river flow was 66 km^3 $year^{-1}$ (based on 1980–2005 data), the annual groundwater abstraction 1.6 km^3 $year^{-1}$ and 8 km^3 $year^{-1}$ was reported to flow into the Aral Sea. This would indicate that if existing demand was fully met, water use efficiency for the basin as a whole would be about 66% with some 20 km^3 $year^{-1}$ going to waste. This in turn would indicate that at the basin level irrigation could be highly efficient if crop demand was fully met. Unfortunately due to various supply constraints, including shortages of water in some years, restricted canal carrying capacities, poor maintenance of infrastructure and weak water management the actual amount abstracted falls short of meeting demand in most years.

Using available abstraction data the, modelled mean average actual total consumptive use by the crops plus water for industrial/domestic use is 30 km^3 which is 25% less than the theoretical requirements. This indicates that the overall efficiency of the basin is about 50%. In practice, more water than the theoretical consumptive requirements needs to be abstracted to satisfy both distribution losses and other water user inefficiencies. These occur due to pipe leakage, losses due to seepage from canals, in field seepage losses as well as losses through the drainage systems. Such losses are notoriously difficult to quantify. For example, water apparently lost to shallow groundwater beneath irrigated fields may be used later by the crops extracting water by capillarity from the shallow groundwater, or the groundwater may seep laterally to adjacent lower lying areas where it can provide a subsurface source of irrigation water. Equally, however, groundwater may seep laterally into adjacent fallow land and cause salinisation.

This all indicates that with the current cropping system and municipal consumption, if crop demand were to be fully met the system could at best be 66% efficient, but in reality it is much closer to 50%. Most of the "lost" water ends up in lakes in large depressions, in solonchaks and in wetlands scattered throughout the basin. There are both anecdotal stories and factual evidence that considerably more water is abstracted than is reported, in some cases up to 30% more. The estimate of 50% efficiency is therefore likely to be low with the real value being between the two estimates. It is important that the quality of the recorded abstraction data is improved since without reliable data it is impossible to optimise water use efficiency and this can lead to the wrong management decisions being made.

If the official abstraction data is correct then the findings from the water resource management model, AMU-WEAP_100, indicate that infrastructure capacity constraints (e.g. poorly maintained canals and pump stations) prevent even the current crop water demand from being satisfied in an average flow year, a situation that is aggravated further in low flow years. Such shortfalls do not automatically result in un-cropped land, and in many areas with shallow groundwater, in drought years, cotton may rely on one irrigation at the start of the season and then use shallow groundwater to meet demand later on in the season. In other areas farmers may switch from cultivating cotton to wheat which can produce a crop with just one irrigation at the start of the season. Nevertheless in our experience such shortfalls are not shared equally but are suffered much more by the irrigators at the ends of the system (e.g. Khorezm and or Karakalpaktstan).

By 2085, in addition to increased evaporation, it is expected that the flows in the rivers will have decreased appreciably. The runoff modelling studies indicate that a mean reduction in flow of 15% is highly likely, although this could rise to over 30% in dry years. Chapter 5 (Figure 5.6) depicts the modelled shortfalls in the water resources available to meet the evaporative demand of the current irrigated command area of the Amu Darya Basin under a number of possible 2085 scenarios. One hundred years of stochastically generated runoff data for the reference period (notionally 1975), has helped identify the range of discharges that might statistically be expected in a given year (Chapter 3 and Chapter 5). The effect that an increase in summer temperatures of +5°C by 2085 will have on the ability of the river to meet crop water requirements and the adverse effects of climate change induced reduced river flows on the shortfalls is considered for 3 scenarios: a modest reduction in river flow of 10%, a medium reduction of 20%, and a large reduction of 30%, the latter only being likely in the drier years when glacial flow plays a more significant role.

Figure 8.5 illustrates the reduction in irrigated area that will result from these shortages, whilst Figure 8.6 presents the probability plots showing the effect of global warming with a medium climate change scenario temperature outcome of +5.5°C and reduced river flows in the Amu Darya river.

The results all indicate that we can expect severe water shortages in the Amu Darya Basin by 2085 and that these will become progressively worse in the foreseeable future. Table 8.1 summarises the shortfall in irrigated areas by 2085 that

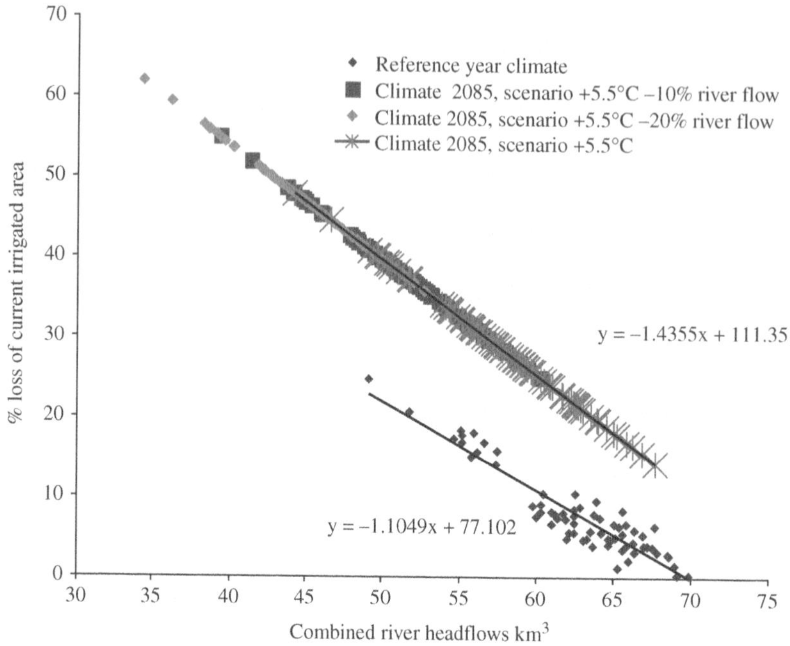

Figure 8.5. The effect of global warming with a medium global warming temperature outcome of +5.5° C, and reduced river flows on the loss of irrigable area in the Amu Darya river basin. (Each point represents the annual flow from within the 100 years of discharge data.)

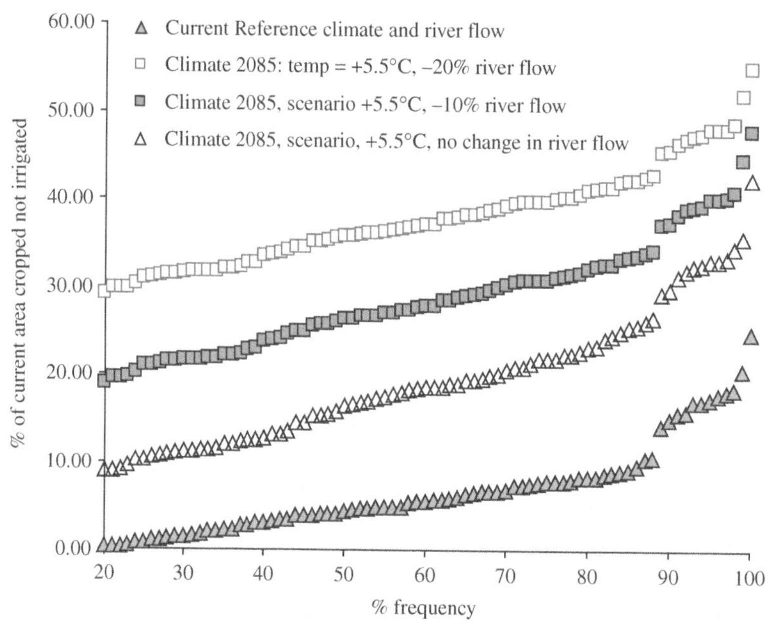

Figure 8.6. Probability plot showing the effect of global warming with a medium GCM climate scenario outcome of +5.5° C and reduced river flows on the Amu Darya.

will arise from a 1:20 year drought for the different modelled scenarios. It should be noted that even under the climate for the reference period (1975), with present water use efficiencies, up to 17% of the irrigated area cannot be irrigated during a 1:20 year drought. By 2085 if river flows remained the same the mean increase in temperature of 5° C would result in a 33% shortfall; but if river flows were to decline as expected by 10 to 20% then the irrigable area will need to decrease by 40% and 48% respectively unless much more efficient use is made of the water. It is not possible to predict with any certainty which of these future scenarios will occur, but even the most optimistic indicate that there is likely to be about 20% less water.

Table 8.1. Proportion of command area of the Amu Darya Basin that could not be irrigated in the 1:20 drought year under the mean 2085 temperature scenario and reduced river flows assuming no operational constraints on abstraction.

Scenario	% command area not irrigable
1975 (notional) Reference period. (Mean average flow)	17
2085 +5.5°C Present River flow Unchanged	33
2085 +5.5°C Present River flow –10%	40
2085 +5.5°C Present River flow –20%	48

The only way that the adverse effects on the irrigated sector can be at least partially mitigated is by improving irrigation water use efficiencies which are presently low, in many areas being only 40 to 50% efficient. Irrigation efficiencies of 60% would be regarded as being very efficient, however, and are not easily achievable at the basin level. The conclusion is therefore inevitable that, even with a considerable improvement in water use efficiency, it is likely that large tracts of land will have to be taken out of irrigation I n the foreseeable future.

8.3 THE ROLE OF EXISTING RESERVOIRS IN PROVIDING A RELIABLE WATER SUPPLY

8.3.1 Fluctuations in annual precipitation and the role of reservoirs in ameliorating shortfalls in water supply

Although the annual sustainable glacial melt plays a major part in dampening fluctuations in discharge between wet and dry years there is still a sizable variation in annual discharge, and in dry years there are water shortages throughout the basin, with the situation being most severe in the lower parts of the basin. The modelling studies showed that the storage capacity of the basin's reservoirs are in general sufficient to provide intra-seasonal (within season) water storage for municipal and industrial use (excluding power supply), but unfortunately they have insufficient storage capacity to provide significant inter-seasonal storage (between seasons) to satisfy irrigation needs in dry years. They can therefore only be regarded as being annual operational reservoirs used to match monthly supply to seasonal demand. Even the largest, the Nurek reservoir, which has an active storage capacity estimated at 4.7 km^3, is unable to provide inter-seasonal storage for dry years because of the need to maintain head at a high level for reliable hydro-power.

8.3.2 The threat posed by siltation on reservoir storage and the water resource management of the basin

The waters of the Amu Darya and its tributaries contain a very high silt load, with 4 g l^{-1} of silt being common during periods of high flow: at times of land/mud slides in the upper basin silt loads can be extremely high. Unfortunately information on the siltation rates of the reservoirs is limited. For example, the Nurek Reservoir on the Vaksh, which on completion in 1972 had a total storage capacity of 10.5 km^3 (an active storage of 4.5 km^3), had decreased by 1989 to 8.66 km^3 (Sherman and Rafikov, 1992), indicating a loss of storage of about 1% per year. This implies an estimated loss by 2009 of 35% of its total volume and the total loss of all effective storage by 2074. This is likely to be a conservative estimate, however, since in 1976 work commenced upstream on the huge Rogun Dam. Construction halted in 1993, when the rock fill dam which had attained a height of 60 m was destroyed by a very large flood. Most of the dam and the impounded sediment entered and added to the sediment load within the Nurek reservoir.

Since the reservoirs do not provide inter-seasonal storage the high sedimentation rates cannot adversely affect inter-seasonal water security. Sedimentation does, however, have a major impact on the storage capacities of the many annual storage and operational reservoirs on the tributaries of rivers and canals that are used to capture and store snow melt water, both for domestic water supply throughout the year and to help provide water to meet peak crop water requirements in mid-summer. Concerning the largest flow-regulating reservoir, the Nurek, unless the Rogun Dam is constructed upstream it will become completely ineffectual for regulating monthly flow. What is clear is that due to the high silt load all of the on-stream reservoirs are silting up at an alarming rate, with several already becoming ineffectual due to siltation. Siltation will significantly and adversely affect the ability of the 35+ reservoirs in the Amu Darya basin to regulate seasonal demand for both irrigation and municipal water supply.

8.4 RISKS AND OPPORTUNITIES FROM IRRIGATION EXPANSION IN THE STATES OF THE UPPER AMU DARYA BASIN

8.4.1 Irrigation development

Future development of water use by the riparian countries of Afghanistan and Tajikistan poses both a current and a future threat to the availability of water for irrigated agriculture in the downstream States. Nearly all effective runoff in the Amu Darya basin rises in the mountains of Afghanistan and Tajikistan, with 13% to 19% being generated within Afghanistan, but these two states only consume a small fraction of this. Both countries therefore consider that they have a legitimate right to more of their 'own' water. In Tajikistan topography limits the scope for further development of economically viable irrigation, but economic viability has not necessarily prevented such developments in the past since there are already some 7000 ha of irrigation necessitating an uneconomic pumped lift of over 200 m (Tajikistan *Ministry of Water Resources*). With electricity prices only being notional and falling far short of covering the generating costs, high lift irrigation may appear superficially to be commercially viable when it is actually not economically viable. Electricity has as an opportunity cost which is related to international gas and oil prices and if these are used in the calculation of costs, high lift irrigation does not appear viable. Hence, there is even an economic case for considering implementing a small reduction in irrigated area, although this would be politically difficult in a country with rapid population growth of 3% per annum, and suffering from food shortages. Nevertheless the Tajikistan Government has a stated desire to develop a lot more land for irrigation; but the steep nature of the non-irrigated land in the country restricts the potential for further development of economically viable schemes. Furthermore, the precarious nature of the Tajik economy probably means that in the short to medium term any increases in irrigated area within the Amu Darya basin in the country are likely to be small and have a relatively minor effect on the availability of water downstream. Any significant further irrigation development in Tajikistan would most likely be in the catchment on the Zerafshan river, once a tributary of the Amu Darya, where there are aspirations to develop a further 100,000 ha of irrigation, ,although the economic viability of this is again questionable. If this were to be developed it would inevitably lead to a corresponding reduction in the area of irrigated land in Samarkand province, an outcome that Uzbekistan would be unlikely to welcome.

Acquiring reliable data on the irrigated area in the Amu Darya Basin in Afghanistan is extremely difficult, but this is not surprising given the years of conflict. Older Government statistics stated that there were some 240,000 ha of irrigated land in the Amu Darya Basin, while Ahmed and Wasiq (2004) estimated 385,000 ha. Part of the discrepancy probably arises from the fact that a number of tributaries no longer reach the Amu Darya due to abstraction for irrigation. The largest estimates of up to 1 million hectares of irrigation may be the result of including both these disconnected sub-catchments and the extensive areas of flood irrigation on the alluvial fans adjacent to the Amu Darya; these flood in years of high discharge but remain dry during years of low flow. What is clear is that there is potential to abstract appreciably more water for irrigation than is currently the case. It is generally accepted that some 10 to 15% of the irrigated command area in Afghanistan is currently not irrigated due to the poor condition of the infrastructure, and there are ongoing projects to rehabilitate these areas. When these programmes are completed they will bring about 50,000 ha back into irrigated agriculture, which in turn will need to be compensated for by an equivalent reduction in irrigated area in the lower parts of the basin.

It was considered that there is economic potential to develop an additional 58,000 hectares in Afghanistan. This will involve a 20–30 m lift from the Amu Darya combined with a 40 km long lined transfer canal to convey the water through the (Barkan) sands. However, much of the non-irrigated lower alluvial fan of the Balkh river could be irrigated, involving an area of some 200,000 ha. Given the current social, economic and political situation in Afghanistan such developments are unlikely in the near future, particularly given that such expansion would require a corresponding reduction in the irrigated area in the downstream States: hence finding international funding for constructing and operating such schemes is unlikely, although not inconceivable. It should be noted, however, that such schemes would be far more economically viable than the existing schemes of the Karshi-Kashkadarya canal in Uzbekistan which involve pumped lifts in excess of 150 m.

It is concluded that in Afghanistan, development/redevelopment could increase the existing irrigated area by up to 100,000 ha in the foreseeable future, but since all the water in the Amu Darya is effectively abstracted, this would inevitably require a corresponding reduction in area in the Lower Riparian States.

8.4.2 Development of hydropower capacity and irrigation

The 300 m high Nurek Dam in Tajikistan has a total storage capacity of 10.5 km^3 but to ensure effective generating head for optimum hydropower output only 4.5 km^3 of live storage is used, resulting in an annual maximum drawdown of about 60 m. In the past, operational practices endeavoured to ensure the reservoir was full at the start of spring in order to provide the maximum amount of water for downstream irrigation in Turkmenistan and Uzbekistan; in recent years, however, there has been a tendency to release more water in winter to meet Tajikistan's own winter power demands, particularly during very cold winters, thus reducing water availability for downstream irrigation.

In both the upstream states of Tajikistan and Afghanistan there are sites where large reservoirs could be developed in the basin if an economic case could be made. During the era of the Soviet Union the 335 m high Rogun Dam was under construction in Tajikistan in order to provide 3500 MW of power, improve reliability of flows for irrigation in Uzbekistan and Turkmenistan, and prolong the life of the Nurek reservoir by trapping sediment upstream in the new reservoir. Unfortunately this was destroyed in 1993 during construction, and further work was halted. Studies have also been made of a number of other viable sites, including an even larger dam at Dashtijum situated between Afghanistan and Tajikistan on the Pyanj. The economic justification of these ambitious developments depends on the willingness, political stability and capacity of the surrounding states of Pakistan, Afghanistan, Uzbekistan and possibly China to purchase competitively priced power. With fuel prices predicted to rise, these projects are likely to become economically feasible and will be realised at some time in the foreseeable future if practical and political issues can be resolved. The only loss of water resulting from construction would be due to increased evaporation from the surfaces of the reservoirs, with some 6000 ha of reservoir surface equating to a similar loss of irrigated area downstream. Since effectively all water is presently abstracted from the river, the period of reservoir filling will result in a substantial reduction in the availability of water for the downstream riparian states, since each reservoir will require 10+ km^3 of water to fill (equivalent to over 70,000 ha of irrigated agriculture). If constructed it is anticipated that these dams would be operated for the benefit of the nations in which the reservoirs are situated in order to meet their national demands for power, and not for the benefit of downstream riparian States where there is a demand for water for irrigation which is out of phase with power demands. Such developments will undoubtedly add to the water disputes between the Riparian States.

There is considerable potential in the region to develop less grandiose hydroelectric projects on many of the smaller tributaries which could operate on 'run of river' principles. These are likely to be more environmentally acceptable, have lower energy production costs and be more acceptable to downstream states since they will not impound water for use in winter. The problem, however, is that these would only operate in summer when there is adequate flow , but the peak demand for energy in these upstream states is in winter. They are therefore very limited in their ability to mitigate power shortfalls.

At the present time Afghanistan uses very little of the basin's water resources and is not signatory to any binding water-sharing agreement with the other riparian states. Developments in Afghanistan are being held back by internal strife but it is reasonable to expect that in the future it has the right to use a fair share of the basin's water resources if its economy is to develop. It is therefore not a question of 'if' Afghanistan will develop its water resources in future but 'when'. It is impossible to predict the final level of development, which will depend on many interlinked factors, but it is essential that the future water resource plans of the riparian States downstream are sufficiently robust to accommodate the expectation that Afghanistan will want to use its perceived share of the water.

8.5 ECONOMIC ISSUES: NATIONAL AND AGRICULTURAL ECONOMIC AND COSTS IF THE RIPARIAN STATES MOVE TO A MARKET ORIENTATED ECONOMY

8.5.1 Pumping costs

Large parts of the irrigated area rely on pumping, often involving very high lifts (see for example Table 8.2 for the lift heights to irrigated areas in Tajikistan). The cost of power means that pumping to these heights is very expensive. In Tajikistan pumping water above 50 m results in over half the land in the command area having a negative economic margin, whilst above 100 m nearly all irrigated land becomes economically non-viable. In Uzbekistan the pump stations along the Karshi-Kashkadarya canal lift water up some 170 m in order to irrigate an estimated 390,000 ha of land. According to Bucknall *et al.* (2003) when the costs of operation and maintenance are taken into account, at an assumed energy price of US\$ 0.05 kWh^{-1}, much of the agriculture of this irrigated massif is not economically viable and is only sustainable and financially viable with vast state subsides.

The Amu Bukhara canal lifts water 30 m, 68 m, and finally to 90 m in order to irrigate a total of some 270,000 ha. The area irrigated using a 30 m lift appeared to economically viable at the assumed energy price of US\$ 0.05 kWh^{-1} given that the capital costs can be considered as being sunk costs. For the higher lifts, however, when the full costs of rehabilitation as well as operation are included, farmers' profit margins become very small. For example, the gross economic farm gate margins for cotton production in Uzbekistan, excluding the costs of water, were estimated to be about US\$ 482 ha^{-1} $year^{-1}$ (assuming an average world cotton price of US\$ 1300 a tonne of lint and a yield of 2.2 tonne ha^{-1} of lint cotton) (Bucknall *et al.*, 2003). The costs of raising water to different heights at an electricity cost of US\$ 0.05 kWh^{-1} are shown in Table 8.3. When these costs are subtracted, the profit from lifting water up 50 to 100 m results in a net profit margin of 175 to 214 US\$ ha^{-1}, whilst for lifts between 100 to 150 m the profit margin falls to 46 to 175 US\$ ha^{-1}.

Table 8.2. Height of pumped water lift for irrigation in Tajikistan. Ministry of Irrigation and Water Resources.

Pumping Height (m)	Area (×1000 Ha)
50+	247
100–150	52
150–200	45
200–250	5
250–300	2.5
Total	349.5

Table 8.3. Cost of electricity to irrigate 1 ha of cotton in Uzbekistan at different pumping heights.

Lift (m)	Pumping Costs* (Electricity cost US$ 0.04 kWh^{-1})	Pumping Cost* (Electricity cost US$ 0.05 kWh^{-1})	Pumping Cost* (Electricity cost US$0.06 kWh^{-1})
50	153	188	223
100	256	307	358
150	359	436	513
200	462	565	668
250	565	703	843

*Average Pumping Efficiency 80%. Depreciation and O&M 8% of electricity cost of pumping. Operation and maintenance of canal and drainage systems US$50 ha^{-1}. Note current expenditure on O&M is only about US$15 ha^{-1} for irrigation water management which is not enough for sustainable maintenance.

If farmers had to meet the true cost of water, irrigated agriculture would not be financially viable for lifts over 50 m, since even a 10% reduction from the mean yield of 2.2 tonnes ha^{-1}, from whatever cause, would result in farmers losing money. Bucknall *et al.* (2003) came up with a more positive view that irrigation in Uzbekistan could be economically viable with lifts up to 150 m. However, they made the very optimistic assumption that the cost of maintenance for the irrigation and drainage systems was only US$15 ha^{-1} $year^{-1}$. The costs of pump operation and maintenance costs were also excluded (except power). The authors consider that US$15 ha^{-1} $year^{-1}$ is wholly inadequate for sustainable management of large irrigation projects, especially those relying on rivers containing high sediment loads, and consider values of US$ 50–6015 ha^{-1} $year^{-1}$ to be far more realistic.

A more positive view suggests that irrigation in Uzbekistan could be economically viable with lifts up to 150 m. However, the authors made the very optimistic assumption that the cost of maintenance for the irrigation and drainage systems was only US$15 ha^{-1} $year^{-1}$. The costs of pump operation and maintenance costs were also excluded (except for power). The authors of this report consider that US$15 ha^{-1} $year^{-1}$ is wholly inadequate for the sustainable management of large irrigation projects, especially those relying on rivers containing high sediment loads, and consider values of US$ 50–150/ $year^{-1}$ to be far more realistic.

By world industrial energy price norms (see Figure 8.7), the prices of electricity in Uzbekistan are low and do not reflect the true costs of generation and distribution. Energy prices are expected to rise steeply over the next 30 years. This will inflict severe economic strains on schemes involving high lifts. They will, frankly, be uneconomic and will impose a continual drain on the national economy. To close them in their entirety is quite unthinkable, however, since hundreds of thousands of people depend on them for their survival. It is concluded therefore that economic viability is a risk which poses a major threat for their long term viability, particularly when considered alongside the expected reduction in the availability of water.

8.5.2 Rehabilitation and maintenance costs

There are an estimated 3.4 million ha of irrigated agriculture in the basin. If the irrigated agriculture is to remain sustainable nearly all the systems and management structures need to be rehabilitated, including land grading, rehabilitation of canals and existing drainage systems, the installation of new drainage systems in salt-affected lands, improved monitoring and management systems and the rehabilitation of many of the control structures and pumping stations. Experience indicates that rehabilitation costs are likely to cost between 1500 to 3000 US$ ha^{-1} depending on the severity of the problems in the different areas, which gives a mean total cost of about US$8 billion. This indicates that in addition to annual operation and maintenance costs, if irrigation is to be rehabilitated over a 20-year period an estimated US$400 million a year needs to be found. Existing plans do not approach this level of expenditure which suggests that overall the irrigation infrastructure is still deteriorating.

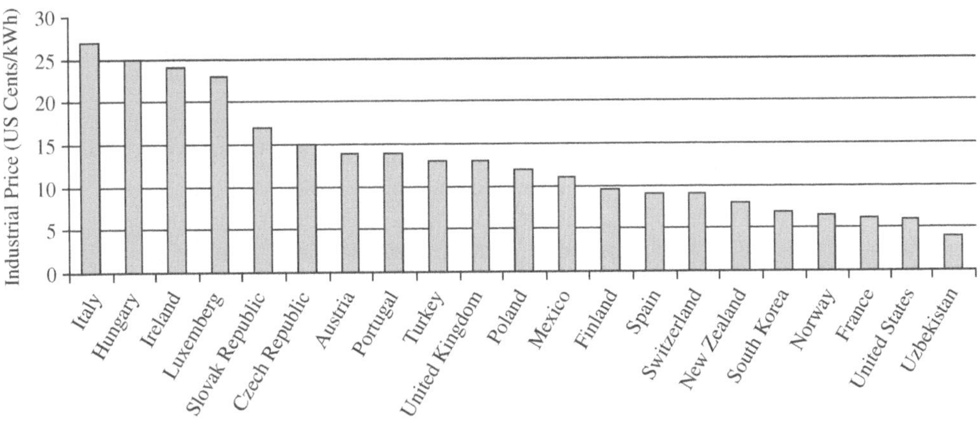

Figure 8.7. World industrial energy prices (taken from: Key World Energy Statistics, 2008. International Energy Agency, Paris).

8.6 CATASTROPHIC AVALANCHE, EARTHQUAKE AND LAND SLIP IN THE UPPER BASIN

The High Pamirs are young in geological terms and the area is seismically highly active, hence landslips, mudslides and high speed landslides from collapsing terminal glacial moraines are relatively common, as are surging glaciers. Although many result in little more than very high sediment loads, others in recent history have caused major catastrophes or near-catastrophes. In Tajikistan major events occurred in 1969, 1970, 1985, 1988, 1993, 1998, 2005. The 1993 event destroyed the 60 m high dam at the Rogun Dam construction site together with many houses and a loss of many lives. In 1911 a landslip created the 600 m high debris dam obliterating the Usoi village on the Murgab River. It holds back an approximate 17 km^3 of water in the Saraz lake. If the debris dam were to fail, by any one of a number of likely mechanisms the ensuing flood would endanger hundreds of thousands of people's lives in the Murgab, Pyanj and Amu Darya valleys. The above shows that the area has a long history of problems resulting from landslides that have caused severe flooding and loss of life and there is the potential for a truly cataclysmic event. Although, such events are difficult to anticipate, it is clear that they need to be considered when planning both water resource infrastructure design and the location of settlements to minimise the potential risks.

8.7 SALINITY

The second biggest threat to irrigation in the basin after reduced river flows resulting from climate change is salinisation of the irrigated fields. The amount of salt entering the middle and lower basin can be calculated from the water quality data at Termez. The mean annual flow of water is 65 km^3 year^{-1} containing some 36 million tonnes of salts. Although much of the salts will precipitate in the irrigated soils as carbonates and sulphates, 8 million tonnes per year will remain as soluble sodium chloride. At the Samanbay gauge in the middle delta, due to abstraction the mean annual flow drops to 7.4 km^3 per year, and contains an average 2.6 million Tonnes per year of sodium chloride. Hence 5.4 million Tonnes of sodium chloride are accumulating annually in the irrigated lands between the two gauges. This would require a minimum of 1.5 km^3 year^{-1} of saline drainage water to be removed from the middle catchment in order to maintain an effective salt balance. In addition much of the salt in the water passing the Samanbay gauge will also be deposited in the irrigated soils of the lower delta. This explains why much of the irrigated area in the middle and lower basin has major problems with land salinisation.

In Turkmenistan there are an estimated 1.6 million hectares of land affected by salinity to some degree, and in Uzbekistan another 2.1 million ha. To maintain a salt equilibrium in the irrigated soils of the basin as a whole we estimate that some 5 km^3 year^{-1} of water needs to be allocated.

Salinity increases annually with large areas of land going out of production, Uzbekistan alone loses some 20,000 ha year^{-1} of irrigated land due to salinisation. Unfortunately, climate change is expected to reduce river flows and this will be associated with increased water salinity, which in turn leads to accelerated rates of salinisation in the fields. For example, Work Programme WP2 reported a salinity of 700 mg l^{-1} in the high summer flows at the entrance to the Tuyamuyun reservoir at the head of the delta, but during the winter when flow is low this rises to around 1400 mg l^{-1}. Although salinity is not expected to rise to this level as a result of climate change, it does emphasis the point that as river flows declines salinity rises. This will be further aggravated by increased evaporative demand which will transport 14% more salt into the fields with the additional required water.

8.8 SOCIAL & POLITICAL RISK

8.8.1 Interstate water risks and the interstate water organisations

The Soviet Union's centrally planned economy developed the policy and financed the huge irrigation and hydropower infrastructure of the Amu Darya basin from Moscow via the Ministry of Land Reclamation and Water Management. In keeping with the times, the issues relating to the development and management of the basin's water resources were mainly seen as being technical in nature or related to internal politics within the Union. The development of the basin's water resources was very successful, providing rapid economic growth and bringing development to the region; unfortunately in so doing it created vast environmental and sustainability problems on an unprecedented scale. Moscow-based planners strategically allocated the water resources of the basin between the riparian states to meet development goals, and in the later years of the Union and after a number of years of severe drought in the mid 1970s the Amu Darya and Syr Darya River Basin Authorities (BVO) were established in 1986. They were charged with managing the water resources of these two great rivers in accordance with a plan provided by the Ministry. This sought to provide 48%, 36%, 15.4% and 0.6%, of the annual water resources of the Amu Darya basin within the USSR to Uzbekistan, Turkmenistan, Tajikistan and Kyrgyzstan respectively, with Uzbekistan and Tajikistan each being allocated 50% of the flow of the Amu Darya determined at the Kerki gauge above the Kara-Kum Canal off-take.

Before the final break up of the Union the declining state of the economy resulted in a fall in the availability of funds for operation and management, at a time when the budget needed to grow to meet the needs of an ageing infrastructure and increasing salinity problems. As a result the operational efficiency of the irrigation systems declined and the stressed nature of the system became more and more apparent, although the drying Aral Sea tended to dominate the news headlines. A number of people began to question the sustainability of the management of the water resources of the basin, with major new schemes being proposed to transfer water to the south from the river Ob in north Russia. During this time the central command and control system remained fully operational in spite of budgetary constraints.

Following the break up of the Union in 1991 all central funds stopped and the responsibility for the sustainable management of the basin's water resources was transferred to the five newly independent nation States which formed an 'Interstate Water Management Coordination Committee (IWMCC)' led by the presidents and Water Ministers of the Riparian States. The ICWC was charged with the short and long term planning and development of water resources in the Amu Darya basin. Since that time various other subsidiary interstate organisations have been established to try to assist the integration of planning and managing of the water resources in the basin. For example the Interstate Council on the Aral Sea (ICAS) was set up to establish polices, oversee inter-sectoral coordination and review projects in the Aral basin, and IFAS, the Interstate Fund for the Aral Sea, was charged with coordination of funds provided by the member states. The two organisations were later merged to become IFAS which is now based in Dushanbe, the capital city of Tajikistan. The IFAS board comprises the Heads of the member states or their deputies. The Amu Darya River Basin Authority was made answerable to the Interstate Committee for the management of the water resources of the basin in accordance with the wishes of the committee. These organisations were meant to provide a sound framework for managing the water resources of the basin. The organisations have had some success but unfortunately the reality is somewhat different from expectations, and their ability to manage the water resources of the basin is constrained by many factors. The issues that limit their effectiveness are discussed below together with an examination of some of threats that must be overcome if the water resources are to be effectively and fairly managed and conflict over water avoided.

The problem of decreased availability of water resources in the basin at a time of increased demand is further compounded by the inability of the new riparian nations to work together to develop sustainable solutions to solve a number of water resource related issues.

The River Basin Authority is charged with managing the water resources of the basin, but in practice it is almost totally emasculated, as it does not have effective access to sufficient reliable hydrological information from all member states to enable it to manage the water resources in accordance with the annual interstate water sharing agreements. Even if this data were available, the low levels of capability and of funding would prevent it from achieving its objectives. For example, it has no legal control over reservoir operation and has insufficient and unreliable data on the hydrology of the upper catchment, river flows, abstraction and return flows, most of which are under the control of national or regional authorities. Unfortunately, as everyone who has worked on the management of the water resources of the basin is aware, this has resulted in widely differing databases containing conflicting information on water resource utilisation which, for many understandable reasons, but particularly because of commercial and political interests, has resulted in reluctance by all organisations to cooperate freely in sharing quality data for the good of all. There is also evidence from independent gauged offtake flow measurements that the official data that is shared between States can be a poor reflection of reality. For instance in Khorezm province, a research team measured canal inflow throughout 2005 and discovered that 37% more water was abstracted than appears in the Official ICWC database. Our own water allocation modelling studies of the delta region also indicate that if a mass water balance is to be achieved between the upstream and downstream gauging stations about a third more water must be abstracted in the delta region than is shown in the official statistics of abstraction.

There may be agreements in place between the Riparian States or even between organisations within a State on data collection and sharing, but these do not necessarily result in the free sharing of reliable data which is so essential for the planning and management of the water resources of the basin. Although the situation may be improving, there is still no reliable, effective and commonly agreed hydrological dataset for the entire basin that the water planners within the Riparian States can trust. Lack of regularly updated, reliable and transparent data of the hydrological resources that interested parties can trust makes it impossible to equitably and transparently manage the water resources, both between and within States. Withholding hydrological data, poor data, and unreliable data can all become tools for a State or organisation to protect its interests against the claims of others, which at worst may contribute to the outbreak of hostilities. A good example of what can happen is the dispute over water in the Middle East: the occupied territories not only give Israel land but access to much-needed water from Syria. Israel withholds hydrological data on water abstraction which makes rational debate on water sharing between the adjoining States impossible and adds to the sources of continuing conflict in the region.

Despite the best efforts of the River Basin Authority (BVO) there is a limit to what can be achieved without adequate reliable hydrological data for effective enforcement of annual ICWC water sharing agreements. It is therefore essential to establish a shared and trusted source of hydrological data that is relied on at times of normal flow, since when water resources become more stressed it will be needed to support conflict resolution.

Very large infringements against the agreed water management strategy, such as unplanned winter releases from the Nurek reservoir for power generation, are detected downstream. These unplanned releases are the direct result of reservoirs in the upper catchment being under the control of the Power Company of Tajikistan whose primary objective is to release water to meet the needs for power in Tajikistan, and not for irrigation in Uzbekistan which was the priority in the past. Hence when there is a national power crisis in Tajikistan, as occurs in very cold winters, water is understandably released to generate power for heating in order to prevent national hardship and unrest; but this leads to less water being available for use in the downstream States for irrigation and thus adds to water stress downstream. This causes considerable tension between the Governments of Tajikistan and Uzbekistan. These issues are resolved on an almost annual bilateral basis. At present the Government of Tajikistan has undertaken to stop unplanned winter water releases in exchange for winter gas supplies from Uzbekistan. Many abstraction infringements are believed to occur throughout the basin but go undetected, however, and hence it is in the interests of infringing parties to keep the monitoring system opaque.

Due to the history of the region, most of the higher levels of expertise in the disciplines required for the effective management of water resources are based in Tashkent from where the basins were managed in Soviet times. It is therefore not surprising that since independence the organisations formed to manage the basin's water resources have tended to gravitate to Tashkent, in particular the Interstate Commission for Water Resources (ICWC). Since independence this organisation has been mainly staffed by Uzbek nationals, who may have, or may be perceived by non-Uzbek stakeholders to have, a bias towards Uzbek interests. As a result there are already signs that the ICWC's influence has diminished, with a number of the States and State water related organisations in the Aral basin ignoring it and 'doing their own thing', and many of the disputes over water allocation and use being resolved on a bilateral basis. If, as anticipated, water supply declines in the future and is unable to meet the growing demand within the riparian states, the existing interstate conflicts of interest are expected to escalate. At these times it will be essential that the interstate organisations responsible for the equitable management of the basin's water resources are seen to be fair and independent of any one state. It is important therefore that in future these interstate organisations are staffed on an inter-state basis, with the leadership being rotated between personnel from the different member states, otherwise their moral authority and hence their effectiveness will be diminished even further.

Afghanistan has never been party to any water sharing agreement or to the existing water allocation practices of the former Soviet States. "Although Afghanistan is a key Amu Darya riparian state the other riparian states have established water management structures, which have not included Afghanistan or recognised its interests, and that regional power politics and antipathy towards cooperation, institutional inertia and self interest, have played a major role in this together with Afghanistan's slow emergence from conflict, and its present limited water demands".

The Central Asian States still follow the general water sharing principles developed during the Soviet era which strongly favour the downstream users of Uzbekistan and Turkmenistan. This is not surprising given the fact that there is a shortage of economically irrigable land in the steep-sided valleys of countries in the upper catchment. In the 18 years since the Aral States became independent and the ICWC was formed, the interstate water sharing mechanisms put in place have proved partially effective in resolving many of the water allocation issues, although there have been serious disputes. Examples include disputes over releases of water from upstream dams in Tajikistan for hydropower generation in winter, rather than saving the water for irrigation in spring for use by downstream states. There have also been disputes over the operation and maintenance of the pump stations located in Turkmenistan but relied on to pump water into the Karshi and Amu-Bukhara canals for use in Uzbekistan. Resolutions of these disputes, however, have tended to be negotiated on a bilateral basis.

In spite of annual agreements at Presidential and Ministerial level for fair sharing of the water resources of the basin, with existing monitoring and management practices the implementation of these agreements is difficult, if not impossible, to police and enforce. Hence, in spite of the ICWC agreed allocation of water at the start of the season, in drier years the irrigation of the lower Delta suffers much more severe water shortages than the middle and upper parts of the basin.

Despite the fact that in most years the water is fully abstracted, all riparian states have independent and ambitious plans for expansion of their irrigation systems, which if carried out would raise tensions between them, with Uzbekistan suffering catastrophic losses in dry years. For example, Turkmenistan is currently increasing its irrigated area by another 0.3 million ha. Since all the water is presently abstracted these political plans are not compatible with existing water sharing agreements.

The UNECE Helsinki 1992 convention on the Protection and Use of Transboundary Water Courses and International Lakes sets out a clear framework for the effective management of water and water resources of a river basin and its fair sharing. Uzbekistan has ratified this convention but none of the other riparian states have, despite international pressure on them to do so. This raises concern over future conflicts of interest and each country's perceived need to optimise its own position.

Uzbekistan is particularly at risk of conflict with its neighbouring states. The water it relies on originates in the upstream states which have large reservoirs and the potential for constructing more. The very large irrigation projects in Uzbekistan that rely on high lift pumping in the middle reaches of the river to provide water to Karshi-Kashkadarya and Bukhara, are supplied by huge pump stations extracting water from across the border in Turkmenistan to meet the demands arising in Uzbekistan.

The Tuyamuyun reservoir complex at the head of the delta has the capacity to regulate flow to the delta region whilst providing command to supply both the left and right bank irrigation canals, yet it is situated within Turkmenistan, and leased to Uzbekistan which utilises much of the water released downstream. It would be technically relatively easy, however to divert more of the water to Turkmenistan by constructing a new left bank canal.

When water shortages arise, issues between States that never existed at times of high river flow enter into the arguments about fairness to strengthen a particular State's claim for more water. One such example is the question of whether the water resources in rivers that historically flowed into the Amu Darya, but because of abstraction no longer connect to it, should be taken into account when considering a riparian States claim for a fair share of the basin's water resources? For example, several rivers in Afghanistan are in this category, but the largest river is the Zerafshan, with a mean annual discharge of 160 $m^3\ s^{-1}$, which originates in Tajikistan and historically passed through Uzbekistan before flowing into the Amu Darya below Bukhara. In recent years, due mainly to the development of irrigation in Uzbekistan, it no longer flows beyond Bukhara. Hence in the Soviet water sharing agreement it was not considered as being part of the overall water budget for sharing of the remaining waters of the Amu Darya and is still not considered a part of the current water sharing agreements. The waters of the Zerafshan on the other hand were not specifically excluded from the agreement. With increased pressure on the use of water resources, riparian States not benefiting from its exclusion have a case for arguing that it should be included in the allocation of the Amu Darya basin's water resources, as it is naturally part of the basin in the same way that the Pyandj and Vakhsh rivers are. This approach would indicate that Uzbekistan has a larger share of the basin's water resources than is presently recognised. Each State rationalises its own arguments in support of its claim for a given share of water, albeit on a selective basis, with each state claiming to hold the 'moral high ground'. Resolving such claims fairly and justly is very difficult and is most easily achieved within an international legal framework such as the Helsinki 1992 convention.

Is there a solution to managing increased demand for the water resources by the States in the upper basin and minimising the threats to Uzbekistan, Turkmenistan? Hydropower generation from existing and possible future dams, because of the high winter power demand, results in release of stored water during winter rather than being stored to meet the spring and summer demands of irrigation for downstream users. As pointed out this is already creating conflict between Uzbekistan and Tajikistan, which has been temporarily resolved by Uzbekistan compensating Tajikistan for lost generation capacity if water is stored to meet Uzbekistan farmers' needs in summer. In 2007 hydro-electricity prices were US\$0.02–3 kWh^{-1} whilst thermally generated power was US\$0.045 kWh^{-1}, but this disparity could widen in the future. Clearly the economic opportunity value of hydropower for the Tajiks is the same as of thermally generated power. The Tajiks would therefore have some justification for expecting that their lost generating capacity is compensated at the same opportunity cost as thermally generated power. Given the growing demand for power in the region, particularly in winter, the sale price of hydropower can be expected to rise considerably and approach that of thermal power. This, coupled with the possibility of Tajikistan and Afghanistan constructing new large reservoirs for hydropower generation, is expected to create even greater conflicts of interest with existing downstream users, particularly in dry years. In Soviet times the principal that was applied in the basin was 'do no harm to others, otherwise pay compensation' although the residents of the peripheral lands around the Aral Sea might dispute that this guideline was applied. This appears to be a sound moral basis for agreement, however, and it suggests that an interstate agreement should be signed based upon this principle. The philosophy of such an approach is sound, particularly if you are a tail-end user. But for an upstream Nation State where the water originates which consumes only a relatively small fraction of the water resources, it is possible to argue that all that is required is a fair share, with fair being defined as an 'equal' share of the resource that originate within the country.

On a global scale, history indicates that in spite of both bilateral and in some cases international agreements on water sharing, upstream States nearly always take priority in meeting their irrigation aspirations at the expense of downstream users. Experience of water conflicts elsewhere indicates that resolution in a time of crisis is very dependent on the similarities and differences between the people of the member states, the quality of the relationships between the member states and their diplomatic bargaining positions, i.e. economic power, military muscle, economic interdependence, common languages, and ethnic, social and religious similarities.

Unfortunately considerable tensions already exist between the newly developed States as they strive to establish their national identities and develop their own national interests. Despite the fact that in an average flow year all water is fully utilised, and in low flow years there are severe water shortages, each of the riparian States has ambitious plans to increase its area under irrigation.

Given the above issues and the fact that by 2085 climate change is likely to result in water resources failing to meet even existing demand by in excess of 20%, very serious water disputes between nations on the fair allocation of water will occur and will be a major risk to the stability of the region unless management plans and coping strategies are developed now.

8.8.2 Institutional issues

The management systems within the basin were developed to operate in the centrally planned Soviet economy. Prior to the break up of the Soviet Union, the irrigation systems were under strain due to both spiralling costs associated with the maintenance of an ageing infrastructure and the need to control a growing salinity problem. This was matched by a steady decline in the maintenance budget at a time when it should have been increasing. At independence 18 years ago the new riparian States inherited a crumbling infrastructure and a severe maintenance deficit. At the break-up, all funding for maintenance from Moscow ceased, with responsibility being transferred to the newly independent states who inherited collapsed economies and inadequate structures for sound governance. It is therefore not surprising that with woefully inadequate budgets and staffing, efficient operation and maintenance struggled to survive and remained ineffective for many years. This further exacerbated the decline of an already decaying irrigation infrastructure.

In recent years budgets for operation and maintenance have not significantly improved in Tajikistan: they have increased appreciably in both Uzbekistan and Turkmenistan, but still remain woefully inadequate in comparison to the backlog of work required and are probably not keeping pace with the rate of decline. If further deterioration of the system is to be prevented, rehabilitation of the estimated 3.4 million ha of irrigated land will require dramatically increased budgets as well as institutional reform.

8.8.3 Institutional capacity in the water sector

The catastrophic collapse of salaries in the public sector after the break-up of the Soviet Union resulted in many of the most capable engineers leaving the public sector altogether for better employment prospects in private companies, leaving a significant skills gap in the water sector. This has been aggravated by the retirement of many of the most able and experienced older staff. The skills shortage has been compounded further by many of the most able university students opting for non engineering topics, with students choosing subjects such as Business Studies, Law, and English with International Studies as they perceive them as offering better career prospects than engineering. The best graduates in engineering look to work in the private sector which is better paid. Attempts have been made in a number of riparian states to address this skills gap. One such approach is to send students on masters courses overseas to study modern water resource management techniques and gain wider experience about how water can be managed more effectively. Unfortunately many of the more able remain to work in the countries where they studied, or return to work in the private sector. As a result of poor pay and conditions many of the staff who remain are poorly motivated. Many have drawn attention to these issues and point out the lack of capability and capacity in the public water sector. The problem is severe in Uzbekistan but is most acute in Tajikistan and Turkmenistan. Unless the shortfall in skills and the problem of institutional reform is addressed it is difficult to see how the threats to the water security can be effectively managed.

8.9 DRINKING WATER

A lack of water in deserts is inevitable but in the irrigation massifs a supply of water is not generally a problem although there are issues of portability. There are however severe problems with infrastructure, with inadequate and ageing distribution systems in poor repair and inadequate operation and maintenance budgets to support sustainable water supply. There are also problems of poor drinking water quality in the lower delta. As is the case with irrigation infrastructure, there is a huge backlog of maintenance which technically can generally easily be resolved given adequate funding.

8.10 REFERENCES

Ahmad, M. and Wasiq, M. (2004) Water Resource Development in Northern Afghanistan and its implications for the Amu Darya Basin. World Bank Working Paper No. 36. World Bank: Washington, DC.

Bedford, D.P. (1997) Climate Sensitivity and water Management in the Upper Amu Darya River Basin. PhD, University of Colorado.

Bucknall, J., Kltchnikovs, I., Lampietti, J., Lundell, M., Scatasta, M. and Thurman, M. (2003) Irrigation in Central Asia, Social, Economic and Environmental Considerations. Europe and Central Asia Region Environment and Social Development. World Bank.

Konovalov, V.G. (2007) Past and prospective Chan ge in State of Central Asian Glaciers. In *Glaciers in Watershed and Global Hydrology*. Workshop, Austria, August 2007, pp. 47–52. International Commission on snow and ice.

Konovalov, V.G. and Schetinnikov, A.S. (1980) Evolution of glaciology in the Pamiro-Alti Mountains and its effect on river runoff. *J. of Glaciology*, **40**(134), 149–159.

Savitskiy, A.G., Schluter, M., Taryannikova, R.V., Agaltseva, N.A. and Chub, V.E. (2008) Current and furure impacts of climate change on river runoff in Central Asia. In *Adaptive and Integrated Water Management*. Paul-Wostl, C., Kabat, P. and Moltegen (eds.). Springer, pp. 323–339.

White, C.J. and Tanton, T.W. (2008) Modelling the future climate of the Aral Basin. JAYHUN final report Deliverable 3.2.b

Sherman, S. and Rafikov (1992) Sedementation of the Nurek reservoir. In *Technology and Engineering, (formally Hydrotechnical Construction)*. Vol. 25, No. 10, pp. 668–673. Springer, New York.

Chapter 9

Managing risk to the water resources of the Amu Darya Basin at a time of climate change

Trevor Tanton

9.1 MANAGING RISK TO THE WATER RESOURCES OF THE AMU DARYA BASIN AT A TIME OF CLIMATE CHANGE

The countries of this desert region are effectively totally dependent for their urban and agricultural water supplies on the rivers supplied by the melting snows of the high mountains of the upper catchments. Hence water deficit for the region resulting from climate change poses a serious dilemma for the Riparian States, since without water the fields will simply revert to desert.

All the risks can be summarised in the phrase 'Significantly less irrigation water will be available at the field level in the foreseeable future'. By 2085 the anticipated increase in crop water requirements (evapotranspiration) will be between 7 and 19% (a mean of 14%) whilst the decline in runoff is predicted to be between 5 and 15%. This will result in an effective overall mean reduction in the irrigated area of between 12 and 33% from that which could be irrigated in the reference year of 1975. In drought years the non-irrigated area could exceed 50% of the existing irrigated area. Therefore plans need to be developed with considerable urgency to minimise the adverse impacts of these changes.

The key issues that need to be addressed if the risks to the water resources of the Amu Darya basin are to be minimised are as follows:

9.1.1 Stop further expansion of irrigation development throughout the basin

With rapidly growing populations, all the governments in the catchment have a stated desire to expand irrigation. This is not possible unless it is at the expense of their fellow riparian states. With the present water management practices there are insufficient water resources available to irrigate the existing command area within the basin in all years. It is therefore a waste of funds to develop new areas since all this will do is spread an overstretched water resource more thinly. The first recommendation must therefore be that there should be no more expansion of the areas of irrigated land. The concept that is locally popular is that the area can be expanded if water usage can be made more efficient. The reality is that given the constraints discussed, water usage has to become a lot more efficient than it currently is just to irrigate the existing area.

9.1.2 Full effective Interstate Co-operation between the riparian states of the Amu Darya basin, including Afghanistan

This is essential for the development of a lasting future water sharing agreement together with a strategy for maintaining it which will remain effective in times of severe water stress. If all riparian states joined Uzbekistan in becoming signatories to the United Nations Helsinki Convention on the Protection and Use of Transboundary Water Courses this could provide a sound legal framework to develop such plans. It is essential that efforts are made to encourage Afghanistan to join the interstate water resource management organisations.

The Interstate Commission for Water Coordination (ICWC) and the Interstate Council on the Aral Sea (ICAS) are the top level committees established to co-ordinate water resource management strategies within the Aral basin, whilst the Amu Darya and Syr Darya River Basin Authorities are charged with implementing the plans and managing the rivers. Given that the Aral Sea is now effectively divided, it could be questioned whether this structure is appropriate since Afghanistan, Turkmenistan and Kazakhstan do not share both river basins. The two river basins are vitally important to Uzbekistan's water supply, while Kyrgyzstan and Tajikistan have only minor shares in the water resources of the second basin. Since water resources are best managed at the river basin level it could be argued that the existing structure is not necessarily the most effective for planning the water resources of the two basins as they are now effectively separate.

A more targeted structure is needed focused more at the river basin level since this is the level at which the resources have to be managed. Two designated Interstate River Basin Water Coordination Committees directly below the ICWC might be more appropriate than ICAS for resolving water disputes and setting plans for each river. The River Basin Authorities would provide technical support to the Interstate River Basin Committees which in turn would be responsible to the Interstate River Basin Water Coordination Committees. For this to be effective, however, the River Basin Authorities would need to increase their capabilities.

It is essential that the riparian states form joint working groups to consider how in 75 years time a 10%, 20% or at worst 30% reduction in water availability might best be managed in order to minimise the adverse socio-economic impacts and ensure the stability of the region. A starting point would probably be through a framework of technical meetings to establish what the implications of such shortfalls mean and how they might be mitigated. Such meetings would raise awareness of the scale of the mitigation measures that will be required and provide a sound basis for effective planning. Water resource management tools such as the AMUWEAP-100 model developed by JAYHUN using the water allocation software package "WEAP" could provide a sound basis for examining the implications of different management options. Once the social and economic implications of these reductions in water resources are fully identified and widely accepted, it becomes possible to develop and implement rational interstate and national coping measures.

Since independence the former Soviet States have worked to establish their own national identities, and in so doing have not surprisingly been developing their own strategies and priorities in order to ensure that their aspirations can be met. The result is that as the pathways of development between the States begin to diverge, cooperation on water sharing issues has tended to decline and tensions between States have increased. Unless this trend is reversed the region will be in for difficult and turbulent times in relation to the sharing of water resources as scarcity increases. Although the need for a national identity is quite understandable, it is essential that this trend towards increasing separation is reversed when it comes to establishing and policing the equitable allocation of water between the riparian States.

To achieve this, agreed and transparent procedures and protocols will need to be put in place for sharing reliable hydrological data. This needs to be at the individual level, the national institutional level and the government level if water allocation is to become fair and effective. A reliable, agreed and shared hydrological database that is regularly updated would form the basis for this. The major internationally-funded WARMIS project to develop a reliable, agreed, shared and freely available hydrological database has failed to meet these needs as it is not readily available to all and data quality is variable. The CAWATER*info* Aral hydrological database goes someway to meeting this need for shared hydrological data but more, and more reliable data needs to be entered. If future conflicts are to be avoided, it is essential that the spirit of fair cooperation in sharing reliable data is driven from the highest political level.

9.1.3 Dealing with dry years

Modelling results indicate that droughts are expected to become more common and their severity to increase whilst the sequences of drought years will also lengthen. With a mean annual flow of 66 km^3 $year^{-1}$, the reservoir capacity needed to provide meaningful inter-seasonal (between seasons) storage would have to retain a minimum of at least 15% of the mean annual flow, or some 9 km^3. A number of factors indicate that this is unlikely to be a viable option for mitigating for dry years. These include:

- The potential sites capable of holding such large volumes of water all lie within Tajikistan and Afghanistan, countries which are not short of water. The main incentive for building dams in these countries is for hydropower generation. It is therefore difficult to see why they should construct reservoirs designed to regulate inter-seasonal irrigation water requirements for use by downstream States. Even if they were interested in building them for this purpose it is difficult to see how they might be financed and operated.
- The economics of building a dam to store sufficient water to provide insurance for dry years that may only occur in say less than one year in three are unattractive.
- Reservoirs designed for hydropower operation and or intra-seasonal water supply are unable to provide inter-seasonal storage.
- The planned Rogun dam in Tajikistan would be able to provide part of this storage if it were constructed to the full height of 320 m. If it is sponsored by commercial finance, however, it is unlikely to be higher than is required to provide a reliable power supply for aluminium processing. If funds were to be provided by international banks to increase the dam's storage capacity, then economic considerations are likely to restrict capacity to intra-season operational supply and not provide for longer term storage.

The only possibility for construction of such large dams would be if they were built on the basis of funding them to provide national security for the downstream States, with political and financial backing for the funding and satisfactory arrangements made to benefit the dam's host State.

The basin has an estimated 15 km^3 of fresh groundwater of which 1.6 $km^3\ year^{-1}$ is abstracted (CAWATER). There is the potential to harness the freshwater in the aquifers underlying many of the irrigated areas. Some of the irrigated areas are underlain by aquifers containing water of high quality; others by aquifers containing saline water. Nevertheless, sizeable lenses of potentially useable water can be found in irrigated areas overlying deep saline aquifers, particularly beneath the higher ground in proximity to the main and secondary canals. These lenses of fresh water commonly overly deep saline aquifers; consequently care is needed when exploiting them. Tube-wells and scavenger wells (i.e. wells designed to skim off the fresh water) could be installed at many sites to supplement water supply in dry periods. Storage underground also has the advantage that evaporative losses are minimal when compared to conventional surface reservoirs and there are no transmission losses. They are recharged by percolation in years of normal or above average flow and thus help to overcome the need to construct large dams. The disadvantage is the capital which would be tied up in tube-wells and power supplies. It is recommended that the total volume of useable water resources that exist in these freshwater lenses be fully investigated and an appraisal made of the role they could play in reducing water shortages in dry years.

9.1.4 Water use efficiency

Many of the risks highlighted in Chapter 8 can be minimised if the abstraction of water can be reduced and water use efficiency improved. At the basin scale we estimate that irrigation efficiency is somewhere between 50 and 60%. This level of efficiency is quite common in large irrigated basins.

Since nearly all consumption in the basin is by irrigation, however, this is where we need to look to save water. It is clear that if 40 to 50% of the available water abstracted for irrigation is being wasted there is significant potential for savings. There are many ways in which water use efficiency could be improved, but there is no one simple solution for all the issues. Some of the more generic problems that need addressing and that could make a significant improvement to water use efficiency are discussed below.

Cotton and particularly rice use excessive amounts of water when compared to spring and winter wheat: switching to less water demanding crops might provide an effective way of saving water to cope with drought, and may even allow for some expansion of the irrigated areas to take place. The difficulty is that when the true costs of irrigation are taken into account, cotton and rice are potentially profitable while irrigation of cereals is not, since the price of wheat is dictated by the extensive wheat lands of the Central Asian Steppes where wheat can be produced much more cheaply under rain fed conditions. It is therefore more sensible in economic terms to grow the more valuable cotton and to buy in wheat.

There is, however, considerable scope in the Amu Darya basin to improve irrigation water use efficiencies since this will help mitigate some, though not all, of the anticipated shortfalls.

Issues of water use efficiency and effective strategies to improve this include:

- Water application efficiency in irrigated fields is generally poor and a very high proportion of the water that reaches the fields is lost as surface runoff excess into the drains at the ends of the furrows. Improved land grading or smoothing is needed to increase water application uniformity and reduce the quantities of surface runoff. In many of the soils, particularly those rich in magnesium and or sodium salts (sodium chloride), gypsum applications are needed to improve water infiltration rates. In other soils plough pans need to be broken up to improve infiltration and thus reduce tail losses.

- Seepage from unlined transfer canals is a well recognised source of water loss. Finding viable economic technical options for significantly reducing seepage from these long canals has proved elusive. The issue needs to be revisited, however, if water use efficiency is to be improved. A better understanding of the amount of water being lost in different parts of these canals may help in identifying economic solutions.
- Water resource management at all levels is weak and is a major cause of inefficient water use. The causes are many and multi-faceted and are often different between projects or at different levels within schemes. Farmers adopt different strategies for ensuring adequate water supply, some of which are legal, e.g. the widespread use of cheap pumps to lift water from the drains into the fields, or the temporary blocking of open drains to provide a secondary source of water, while others are illegal such as hidden off-takes, illegal abstraction at night or offering incentives to the water bailiffs to ensure supply. It water is to be managed efficiently and fairly and excess wastage avoided, improved water management needs to be addressed by:
  - Abandoning the most unproductive land or those lands suffering very high seepage losses.
  - Consolidating irrigated lands into large continuously irrigated blocks and thus avoiding fallow fields within the blocks which actively encourage lateral seepage losses and result in both loss of water from the system and a build-up of salts in the fallow areas.

Drip irrigation may be thought of as being a solution to saving water but for a number of reasons it is not. It is not practical or effective when used with water containing high silt loads nor is it effective for field crops without the provision of substantial State subsidies. Drip irrigation also requires highly motivated and careful field operators to maintain it and prevent it from becoming damaged.

- Much of the water in the field drains in the summer months is of useable quality since much originates from the surfaces of fields as tail-end runoff and hence it can still be used for agriculture. For example even in the highly saline Karakalpakstan region, in the irrigation months of June, July and August when discharge can reach 39 $m^3\ s^{-1}$, the water quality in the left bank drain did not exceed 2.5 g l^{-1} (electrical conductivity ~4dS m^{-1}). The other two major drains in this salt affected area just exceeded this threshold. In most other areas in the upper and middle catchment drain water quality is much higher. As a general practice we would recommend returning all drainage water with an electrical conductivity of less than 4 dS m^{-1} (<2.5 g l^{-1}) directly to the nearest irrigation canal unless it can be returned to the river without further losses. This approach is particularly important for water that has been lifted many metres at very high cost since it only requires a very minor lift to return it back to the canal. It is envisaged that such an approach, if systematically applied, could go a long way to minimising existing losses and will significantly reduce the flow to the depression lakes, many of which are nearly full and in danger of becoming ineffective.

Care however is needed as experience has shown that blending water with salinities greater than a concentration of about 2.5 g l^{-1} with fresher irrigation water does provide an emergency solution to crop water shortages in dry years, but used systematically each year can exacerbate long term salinity problems. If poor quality drainage water (EC >2.5 g l^{-1}) is discharged back into the rivers in appreciable quantities this also leads to a rapid decoration of water quality downstream. Subsurface drains discharging water with salinities in excess of 2.5 g l^{-1} are by world standards considered to be rather efficient at removing salt, and wherever possible water with salinity greater than this should be disposed of elsewhere, e.g. in depression lakes.

9.1.5 Salinity control

Saline soils which do not rely on high lift pumped supply should be provided with subsurface drainage systems as funds permit, but care must be taken in their design to minimise the loss of excess water. In the irrigation season the quality of drainage water from shallow drainage systems is often not much different than the quality of the irrigation water. Once a new drainage system has lowered the high concentrations of salt in the saline soil to satisfactory levels, loss of water in the main irrigation months of June, July and August should be prevented. This can be achieved by raising the water levels in the open collector drains to prevent the drains from functioning. This approach is already adopted in many areas of the basin at times of drought.

9.1.6 Structural reform in the irrigation sector

Serious consideration should be given to how the irrigation system is managed and the management transformed to make it more effective. Although considerable efforts are being made to try to reform the management structure in a number of the Riparian States, there is still a long way to go and it is recognised that in all States there is scope for improvements. Unfortunately all too often vested interests or reluctance to change hinders reform.

Salaries for State Employees are very low. It is therefore not surprising that a significant number of those who are employed in the water sector need to consider how they can increase their income. This can be achieved in a number of ways, not all of which are necessarily legal: those which are, as in the case of second jobs, reduce the commitment that staff are able to make to the management of the irrigation systems. If good young people are to be retained it is essential that staff who perform their duties well are properly rewarded while those who fail are removed from post.

At the behest of donor agencies there has been a move to introduce Water User Associations to manage and maintain the tertiary irrigation infrastructure. Our experience in many parts of the world where such associations have been introduced indicates that they have only a limited impact. This is perhaps not surprising given their lack of technical knowledge and training, and the poor profitability on many of the farms. At worse such schemes can lead to deterioration of the infrastructure. We recommend that if they are to be introduced pilot projects are initiated to develop operational structures that are both effective and sustainable, and do not die when the initial start-up funding ceases.

Finally, the life blood of the riparian States is dependent on the supply of irrigation water. If they are to meet the challenges that the future holds, it is important that the value of irrigation and the importance of qualified staff is recognised to ensure that the brightest students choose to become water engineers.

9.2 CONCLUSIONS

It is easy to identify the problems of the irrigation systems in the Aral basin. It must be recalled, however, that overall the irrigation in the basin has been and is a great success in terms of the improvement to the quality of life that it provides for the millions of people who depend upon it.

In most years, with the present infrastructure and level of water resource management water shortages are experienced to varying degrees in different parts of the basin. It can be concluded from the above that there are many risks that affect the security of the water supply in the region, but in the medium to long-term climate change poses the most serious threat, with all climate change scenarios and models indicating severe water shortages for the region. Water shortages arising as a result of climate change are likely to be exacerbated even more for the lower regions of riparian States located at the end of the basin by further development of water resources in the upper basin. Although the risks cannot be accurately quantified, the fact is that a shortfall is expected by 2085 of between ~12% and 35% (a mean of 23%), but with the 1-in-20 year drought being much more severe (Table 8.1).

Three future scenarios for water availability need to be adopted by the Riparian States. We recommend that consideration is given to a) 10%, b) 20%, and in the worst conceivable case c) 30% less water resources. The States then need to investigate planning and development strategies that could be adopted for coping with these scenarios and minimising the risks. An interstate forum (possibly established under the IWRC) also needs to consider how current plans might be rationalised to create a basis for further planning and implementation in order to ensure that future investment strategies in the water sector are designed to mitigate water shortages.

Although mainly of academic interest, the greatest need is not for more research to refine the estimates of risk. All the indications are that less water will be available and that financial resources need to be directed to developing and implementing cost-effective solutions to managing the risks. Many rehabilitation projects have been identified that will save water. It is important, however, that the most cost-effective are implemented first. We suggest that funding for research should be directed towards establishing the most cost-effective solutions for water saving, and identifying mechanisms that could overcome potential conflicts over the dwindling resources.

IWA Publishing's authorised EU representative for General Product Safety Regulations is Diane D'Arras, 15 rue Duret, 75116 Paris, France, e-mail: safety@iwap.co.uk.

Printed and bound by CPI Group (UK) Ltd, Croydon, CR0 4YY
12/05/2026
02108874-0018